HOW THE COWS TURNED MAD

The publisher gratefully acknowledges the generous contribution to this book provided by the French Ministry of Culture and by the General Endowment of the University of California Press Associates.

MAXIME SCHWARTZ

HOW THE COWS TURNED MAD

Unlocking the Mysteries of Mad Cow Disease

Translated by Edward Schneider

WITH A NEW FOREWORD BY MARION NESTLE

UNIVERSITY OF CALIFORNIA PRESS

Berkeley   Los Angeles   London

Comment les vaches sont devenues folles, by Maxime Schwartz, is published in English translation by arrangement with Éditions Odile Jacob, © Éditions Odile Jacob, Mars 2001

© University of California Press
Berkeley and Los Angeles, California

University of California Press, Ltd.
London, England

© 2003, 2004 by the Regents of the University of California

Library of Congress Cataloging-in-Publication Data

Schwartz, Maxime, 1940–.
  [Comment les vaches sont devenues folles. English]
  How the cows turned mad: unlocking the mysteries of
  mad cow disease / by Maxime Schwartz; translated by
  Edward Schneider; with a new foreword by Marion Nestle.
    p.   cm.
  Includes bibliographical references and index.
  ISBN 0-520-24337-4 (pbk. : alk. paper)
1. Prion diseases—History.
  [DNLM: 1. Prion Diseases—history. WL 300 s3995c 2004] I. Title.
RA644.P93S3913   2004
616.8'3—dc22   2004006489

Manufactured in the United States of America
13 12 11 10 09 08 07 06 05 04
10 9 8 7 6 5 4 3 2 1

CONTENTS

FOREWORD TO THE PAPERBACK EDITION

READERS OF THIS fine book, which so clearly recounts the scientific and social history of "mad cow" disease and related spongiform enceph-alopathies, should not be too surprised by the spread of the bovine form of the disease to North America in 2003. In May of that year, Canadian officials announced the discovery of mad cow disease—for-mally known as bovine spongiform encephalopathy (BSE)—in an eight-year-old Black Angus cow from Alberta. In October, however, the Harvard Center for Risk Analysis downplayed concerns about the disease and reassured the U.S. government and beef industries that the actions they had already taken were "robust against the spread of BSE to animals or humans should it be introduced into this country." Just two months later, the U.S. Department of Agriculture (USDA) in-formed the cattle industry and the public that BSE had been found in a slaughtered Holstein cow from a dairy farm in Washington State.

One case of BSE in North America might appear as an anomaly, but two begin to sound like an epidemic. Anyone reading *How the Cows Turned Mad* will understand why this is so. Mad cow disease is a foodborne illness. Cows travel in herds. Herds eat the same food. The contaminated feed responsible for this disease cannot be kept under lock and key: it gets distributed into the general feed supply for animals. As explained by a panel of international experts early in 2004, the presence of BSE in two cows indicates that mad cow disease is now endemic in North America.

BSE is demonstrably bad for cows, but it is also bad for people. People who eat meat from affected cows can contract a human variant, Creutzfeldt-Jakob disease. BSE also has profound economic implications that affect human welfare. Cows that have BSE—or are suspected of having BSE—must be destroyed, and no country will knowingly import meat from countries where BSE is endemic. Thus, the history of mad cow disease has much to teach us about how our current systems of food production and food safety oversight affect human health and welfare.

This history also has much to teach us about human biology. As Maxime Schwartz warns, "We have not yet beaten The Disease." By "The Disease," he means all of the various manifestations of transmissible spongiform encephalopathies in animals and people. These diseases are well worth our attention. For one thing, they are invariably fatal. For another, they involve riveting biological phenomena, beginning with their apparent cause—the weirdly infectious proteins known as *prions*. Misfolded prions infect by causing normal prions in the nervous system to misfold, thereby destroying brain tissue (hence encephalopathy) and leaving gaping holes (hence spongiform). The Disease in cows is BSE; in sheep it is scrapie; in deer, Chronic Wasting Disease; and in humans, Creutzfeldt-Jakob disease (CJD) or its BSE-related variant (vCJD).

The Disease occurs spontaneously, but rarely, in older animals or humans, but is not infectious in the usual sense; it is not, for example, spread by sneezing. Instead, it is bizarrely disseminated by one or another form of *cannibalism*—people consuming the brains of deceased ancestors, as in the Papua New Guinea version of The Disease, called *kuru*, or cows consuming the brain tissues of affected cows, as in BSE. Moreover, The Disease permits "species jumps," the transmission of infectious prions from one type of animal to another. Most alarming, misfolded prions are relatively indestructible. They are not removed by normal physiological mechanisms for getting rid of bad proteins, nor are they inactivated by common methods of sterilization. Once eaten, they resist destruction by digestive enzymes, stomach acid, and immune systems. Although a great many biological questions about the structure, function, and transmission of prions remain to be answered, one possible scenario—as yet to be confirmed—is this: Cows "caught" BSE by eating the brain tissues of sheep infected with scrapie and people contracted vCJD by eating meat contaminated with brain tissues from cows with BSE.

How this happened and how scientists came to understand The Disease constitute the principal narrative themes of this book. Professor Schwartz tells the story chronologically, beginning with the earliest written descriptions of scrapie in the mid-eighteenth century and ending with estimates of the number of deaths expected from variant Creutzfeldt-Jakob disease 250 years later. Because he pauses at the beginning and the end of each chapter to summarize, and because he writes with brilliant clarity, he makes both the political history and the science of prion diseases accessible to anyone. Readers of this book will be well prepared to understand ongoing discoveries and developments, scientific as well as political, as they occur.

As is now well known, BSE caused an epidemic in British cows several years after the animals began to eat feed contaminated with misfolded prions. To understand how infectious prions got into their feed supply in the first place, one need only consider the enormity of the problem posed by the inedible parts of cows—their offal. In the United States, for example, nearly thirty-six million cows were slaughtered for food in 2002. People do not usually eat cow bones, intestines, ears, hooves, organs, brains, or spinal columns; these and other offal amount to about 40 percent of the animal's weight. Something has to be done to dispose of a mass of offal equivalent to this percentage of the weight of thirty-six million cows each year. If the average cow weighs eight hundred pounds at slaughter, we are talking about 12 *billion*pounds of offal per year, and that just in the United States. The easiest and most remunerative disposal method is to render the leftover parts into meat-and-bone meal, an ashy substance that can be sold for industrial processes (cosmetics, gelatin, and the like) and, because it contains minerals and protein, for animal feed.

Prior to the late 1970s, British rendering processes involved pressure-cooking the offal at high temperatures in the presence of organic solvents. When energy costs rose, rendering plants adopted cheaper methods that omitted solvents and used less heat. The cheaper method killed most bacteria and viruses, but did not inactivate misfolded prions. Prion diseases take a long time to develop and the first cases of BSE did not appear until several years after cows began to eat the inadequately rendered meat-and-bone meal.

Although Professor Schwartz suggests reasons why this explanation may be too simplistic, he finds the results unambiguously catastrophic: BSE in 180,000 British cows, four million cattle destroyed, transmission of the disease to other countries through exported meat-and-bone meal, and, by the end of 2003, about 150 human cases of vCJD, mostly among British young people. Fearing the loss of their industry and livelihoods, cattle and beef producers pressed the British government to assure the public that meat was safe to eat and officials did so, repeatedly. Nevertheless, the British government banned the use of offal in cow feed. It also prohibited the use of older cows (which are more likely to have developed BSE) and mechanically recovered meat (which could be contaminated with nervous system tissue) in human food. The United Kingdom instituted testing systems later adopted by the European Union; member states must now test for BSE in all sick cows over the age of two years and in all cows aged thirty months or over. With the introduction and practice of these rules, the epidemic subsided. Cases of BSE reached peak levels in Britain in 1993 and have declined steadily ever since.

Most other countries did not bother to institute such measures until years later. Their cattle industries insisted that BSE was a uniquely British problem and that preventive measures were unnecessary as long as British meat-and-bone meal was not imported and British beef stayed in Britain. Driving this rationalization, of course, were the cost and inconvenience of finding other ways to dispose of leftover animal parts and the implications of preventive measures: trading partners might think that BSE was present. Canada and the United States did not even take the most basic preventive step—a ban on the use of meat-and-bone meal in cattle feed—until 1997. Only in 2000 did U.S. agencies prohibit the importing of rendered animal products from countries that could not prove that their cattle were BSE-free.

Yet much evidence argued that more stringent precautionary measures were essential. In the United States, numerous shipments of animal by-products from prohibited countries evaded the ban and disappeared into the feed supply, untraceably. Although officials said that most of the material had been used for pet food, this fate could not be verified (and in any case would be unlikely to reassure pet owners). Federal inspectors reported that some feed mills were not preventing meat-and-bone meal from getting into the cattle feed supply, and that some processing plants using mechanical recovery systems were not preventing contamination of meat with brain and nervous system tissues. Although meat-and-bone meal was not supposed to be fed to cattle, it could still be fed to chickens, pigs, pets, or farmed fish, under the optimistic assumption that feed intended for consumption by these animals could not inadvertently get into feed intended for cows.

In 2002, the U.S. General Accounting Office (GAO), an investigative agency of Congress, warned that weaknesses in inspection, testing, and enforcement policies for animal prion diseases needed immediate improvement. U.S. regulatory agencies proposed more stringent rules forbidding the use of brain and nervous system tissue and of offal from older and "downer" cows (those that died before slaughter), but these efforts were blocked by political officials. The cattle and beef industries continued to argue—apparently persuasively—that stricter preventive measures were unnecessary on safety grounds and would impose need-less regulatory barriers and raise production costs.

Even after the emergence of the two BSE cases in 2003, the official position of the U.S. government continued to be that the only cows at risk of BSE were those that ate contaminated meat-and-bone meal prior to the 1997 ban. Because older cows would not be alive much longer, the BSE threat could be expected to disappear along with them. Officials of Canada and the United States must have been relieved to learn that both cows with BSE were old enough to have eaten contam-inated feed before the bans were in place. In addition, when UDSA of-ficials discovered that their affected cow had been imported from Canada, they could continue to claim that American cows were free of BSE. Canada, however, shipped 1.7 million cows to the United States in 2002, and cows are transported freely back and forth across the Canadian and Mexican borders. British experts pointed out that the only reason BSE had not been found in the United States was that U.S. officials were not trying hard to find it. The United States tested just 20,000 cows in 2002, a rate substantially below that of Great Britain, Europe, and especially Japan, which tested *all* cows intended for hu-man consumption.

One consequence of the two BSE cases was to expose aspects of beef production and processing—and their oversight—not generally known to North Americans. Both cows had been tested for BSE almost by ac-cident. The Canadian cow was visibly ill, and the U.S. cow was said to be a downer (an observation strongly disputed by its slaughterer and other witnesses), yet both carcasses went immediately into the feed or food supplies. The Canadian cow was slaughtered at the end of Janu-ary and its head sent to a laboratory for testing; the results were not re-ported until May, by which time its carcass had long been rendered and distributed. The brain of the American cow was shipped to Iowa for testing on December 11. By the time the results came back on De-cember 22, its meat had been ground into hamburger along with the meat of 19 other cows slaughtered that day. Neither country could adequately trace the origin and history of the affected cows, their calves, or the products made from them. Parts from the Canadian cow moved through at least eight feed mills, two farms, and two pet food companies. Hamburger from the American cow and her companions was reported to have gone to several mainland states, Hawaii, and Guam. The USDA was able to find less than half of the herd members shipped from Canada with the affected cow and eventually gave up the search. The cases also exposed the USDA's lack of authority to force packing plants to trace and retrieve meat that might be contaminated.

Despite these problems, USDA officials spent the last weeks of 2003 and first weeks of 2004 issuing repeated reassurances that the 1997 ban on meat-and-bone meal protected the beef supply, that the single U.S. case was an anomaly (particularly because it had come from Canada), and that U.S. meat posed no safety concerns. USDA Secretary Ann Venemen said, "I plan to serve beef for my Christmas dinner. And we remain confident in the safety of our food supply." USDA safety officials said, "The recalled meat represents essentially zero risk to consumers," and U.S. President George W. Bush told a reporter for the *Washington Post,* "As a matter of fact, I ate beef today and I will continue to eat beef."

To anyone following the history of mad cow disease, such statements must surely have elicited feelings of *déjà vu,* as they closely echoed sentiments expressed by British officials in the early 1990s, just as the BSE crisis was accelerating. The British statements undermined public confidence in the government's oversight of food safety. Even in the United States, where people have more trust in government, this single domestic case of mad cow disease raised many uncomfortable questions: Why was meat from a potentially sick cow permitted to enter the food supply at all, let alone before the results of testing became available? Why was this cow's brain tested as part of a routine surveillance program rather than an ongoing safety program? Why were the origins and history of this cow and its meat not readily traceable? Why doesn't the USDA have the authority to close plants that produce unsafe meat or to recall meat suspected of contamination? And why were USDA officials so certain that the meat supply was safe, when they couldn't answer such questions? The U.S. government did not seem to have learned much from the experience in Great Britain. Indeed, food safety advocates viewed the situation as further evidence that government agencies were favoring the commercial interests of powerful industries over the health of the public.

As it turned out, industry interests might have been better served by demanding the strongest possible government regulation of food safety. Canada's beef industry collapsed after its one case was announced, and the U.S. case caused the refusal of American beef imports by nearly 50 countries—threatening $3.5 billion in annual exports—as well as drops in cattle prices and futures, quarantines of U.S. cattle, and widespread layoffs in the beef industry. To protect the industry, government agencies instituted the very policies they had deemed excessive just a few weeks earlier. Despite intense lobbying against such measures by the meatpacking and cattle industries, the USDA banned downer cows from the human food supply, required the return of test results before allowing meat into the food supply, prohibited air injection of cattle before slaughter (a practice that disperses brain tissue), imposed more restrictive controls on meat recovery systems, and ordered the creation of a nationwide tracking system. These rules made sense, but left loopholes troubling to food safety experts. Downer cows could still be fed to chickens, pigs, pets, and farmed fish. Rendered hogs and chickens could still be fed to cows. BSE testing rates, although raised substantially, still fell well below the standards of the European Union. Furthermore, the USDA continued to resist calls for more efficient systems of traceability and recall.

The USDA's hesitancy to take more comprehensive protective actions can be explained by the many forces opposing such measures. The strongest opposition comes from the insistence of the powerful cattle and meatpacking industries that government pass no regulation that might increase production costs or even remotely suggest that meat might not be perfectly safe or healthful. A second opposing force derives from pressures to keep the cost of food as low as possible. Inexpensive food enables the population to be fed adequately (a benefit to society) but it also stimulates sales (a benefit to industry). Cheap food, however, comes at a price—in obesity (low cost stimulates consumption), in fierce competition among food companies to keep costs down, and, as in the BSE situation, in cost-cutting production practices that sometimes cut corners on safety. To maintain an inexpensive food supply, government agencies try to avoid imposing unnecessarily restrictive regulations. Corporate donations to elected officials encourage a relaxed regulatory environment in which it is all too easy to turn a blind eye to safety lapses. Corporate donations to university groups sometimes have similar effects. In the case of BSE, USDA political officials were able to call on the industry-funded Harvard Center for Risk Analysis to reinforce their inaction.

The Harvard Center invoked yet another opposing force—the conflict between views of disease risk that depend on measurement of cases and deaths and those that depend on more subjective perceptions of the degree of danger. Although 180,000 or more cows in Great Britain had BSE, *only* about 150 people had contracted the disease by the end of 2003. In the United States, only one cow (and Canadian at that) had been found with BSE and only one person had variant Creutzfeldt-Jakob disease—and she had lived in Great Britain during the period of greatest exposure to BSE-contaminated meat. On this basis, the beef industry and the Harvard Center maintained that the scientifically proven risk posed by BSE was so small that no further protective measures were needed. This view may be correct on a statistical basis. But The Disease exhibits many characteristics that quite understandably induce public alarm: it is fatal, it cannot be diagnosed until it is too late, and its cause is strange and foreign. Individuals cannot protect themselves against it but must depend on precautions taken by industry and the diligence of government oversight, neither necessarily trustworthy. For all of these reasons, BSE and vCJD rank high on any risk scale of dread and outrage.

Dread and outrage explain why the issue of trust arises so prominently in any discussion of mad cow disease. The absolute risk of contracting vCJD may be low, but it is finite and not at all minimal if *you* or someone you know happen to be one of the victims. Moreover, the cost of failing to institute comprehensive protective measures is incalculably high—not only in human and animal life, but also in the loss of public confidence in food companies and government regulatory agencies. Of course it would be easier to develop policies for prevention of mad cow disease if we knew more about the biology of prions and their clinical behavior. But while we are waiting for the results of further research, it makes sense to implement more careful food safety measures. Such measures are good for everyone. They protect the health of the public and they elicit confidence in government institutions. Although the behavior of many companies might seem to argue otherwise, strong government regulatory controls also benefit the food industry. Regulations—sensibly developed and equitably enforced—establish a level playing field: all companies must follow the same rules. Most of all, they induce trust. *How the Cows Turned Mad* teaches us that thoughtfully designed, diligently followed, and thoroughly enforced food safety regulations will not only protect us against The Disease and other food safety hazards, but are also desirable because they are the right thing to do in today's global food economy.

Marion Nestle
New York, April 2004

PROLOGUE

Thus, we see infection in a new light which cannot fail to be a cause
of concern for humankind—unless, in the course of its evolution over
the centuries, nature has already come upon every opportunity to
produce infectious or contagious diseases, which is highly unlikely.

Louis Pasteur, 1881

There shall be new diseases. That is an inevitable fact. Another fact,
no less inevitable, is that we will never be able to detect them from
the outset. By the time we have some idea of those diseases, they will
already be fully formed—in their adulthood, so to speak. They will
appear like Athena, springing fully armed from the forehead of Zeus.
Charles Nicolle, director of the Institut Pasteur of Tunis and winner of the
1928 Nobel Prize for medicine, 1933

UNKNOWN TO THE public at large until recently, Creutzfeldt-Jakob disease (CJD) is now the subject of daily media attention, especially in
western Europe. What exactly is this disease that seems to pose a threat
to us all? What is the meaning of terms we hear, such as "sporadic," "inherited," "iatrogenic," and "new variant"? What is CJD's relationship
to "mad cow disease"—bovine spongiform encephalopathy (BSE)—
and to the sheep disease known as scrapie? If it can be transmitted from
cattle to humans, why not from sheep to humans? As an infectious
disease, can it be transmitted from human to human? What is the

1

causative agent? It is said to be neither a bacterium nor a virus—so what is it? Can it be detected in infected animals or humans? Can it be eliminated with the right drugs? Is there a vaccine? Why have the cows gone mad? If it is due to their animal-based feed, as some assert, can government prohibitions put an end to the epidemic? And in the meantime, what should we do to protect ourselves? Can we eat beef? Can we drink milk? How many human victims will there be? Dozens? Or hundreds of thousands?

Many of us have asked these questions and more. At the moment, unfortunately, science has only partial answers, and the lack of scientific certainty only fuels anxieties—and wild imaginings.

In France, the "mad cow crisis" grew to incredible proportions toward the end of 2000. It was without doubt a defining event of the final months of the twentieth century. A number of reasons have been suggested for this. Setting aside the isolated incidents that have garnered widespread publicity, there were indeed objective reasons for concern: predictions by epidemiologists that there could ultimately be more than a hundred thousand victims in the United Kingdom alone, the growing number of reported cases of BSE in French cattle, and a scientific finding that suggested the disease could be transmitted through blood transfusions. What was broadly viewed as a genuine panic spread throughout Europe with the late November appearance of BSE cases in Germany and Spain, countries previously thought to be untouched by the disease. The collapse of confidence in products of bovine origin spread to all agricultural products. People no longer knew what they could safely eat—and the economic and political consequences are well known.

The confusion was exacerbated by a parallel crisis: the transmission of Creutzfeldt-Jakob disease to children who had been treated with human growth hormone. That tragedy surfaced in 1985, when the first cases were identified in the United States. And, sadly, it has continued, especially in France, where new cases come to light each year. The public is asking how physician-administered treatments to make these children grow could infect them with a lethal disease.

Creutzfeldt-Jakob disease is frightening because it is always fatal, because it involves the central nervous system and hence the consciousness—the very personality—of its victims, and because we all share a vague fear of contracting it at some point from one source or another. Lest we fall prey to obsessive fear, we must rationally assess the often alarmist information that comes our way. Thus, we need to know more about this mysterious illness—"The Disease"—and must try to understand where it comes from and how it spreads.

When we delve into the origins of this worrisome illness and into the growth of knowledge about it, we find a sort of detective story with its roots in the distant past. It was already lurking in the eighteenth century. First identified among English sheep, The Disease killed all it touched. It was like a criminal, donning ever-changing disguises to elude its pursuers: We have been on its trail for three centuries. That trail begins in Britain, and continues in many other countries such as France, Germany, the United States, Switzerland, Austria, Israel, Australia, and two islands at opposite ends of the Earth—Iceland and New Guinea. Our hunt has made use of the latest scientific advances as they emerged, but its route has also led us through Stone Age civilizations.

Many times, almost as if it sensed that it was about to be found out, The Disease has counterattacked, claiming many victims and spreading fear.

Have we finally tracked it down? Many people think so, but others are doubtful. We shall see.

1

THE SHEEP ARE STRANGELY DIZZY

THE EIGHTEENTH CENTURY, which was to conclude with the American and French Revolutions, was also the Age of Enlightenment. The conviction was growing that scientific progress was intended to enable humans to control the world around us: Had we not learned to control thunderbolts, thanks to the lightning rod? This was the era of Linnaeus, Buffon, and Diderot, and we began cataloguing nature's riches and seeking to employ them in a more rational way.

This approach was seen particularly in the areas of agriculture and animal husbandry. Improved productivity was the order of the day. Landowners organized and agricultural societies and academies were founded, where questions of farming were discussed and where news and information were exchanged and documents published. In England, the enclosure policy was broadly implemented, evicting small-scale farmers to the benefit of big landowners and providing the latter with the resources for long-term investment. In the sphere of animal husbandry, major efforts were made to improve feed and conditions, and to select the most productive breeds.

Sheep farming was the first beneficiary of this modernization because wool production was a major industry not only in England but throughout Europe. It is estimated that a quarter of the English population was involved in wool production or the wool trade in one way or another. And that sector was to remain important. Toward the end of the nineteenth century, an eminent French veterinarian, while noting that sheep farming was on the rise for purposes of meat production, wrote this:

> Wool is among the pillars of the well-being of modern-day societies. We may thus venture to say that the people who produce the most wool will be the richest and perhaps the most powerful. For more than fifty years, wool production has rained showers of gold upon Europe: For Germany and Russia, it has provided hitherto unknown material well-being and the hope of future prosperity. As for England, is it not its countless merinos that must be deemed accountable for the wealth of its colonies and the magnificence of its trade beyond compare?[1]

Given the care lavished upon these wool-bearing creatures, is it any surprise that the diseases that could affect them were also the object of attention? They had to be catalogued in the hope of being able to conquer them. Thus, the existence of The Disease was first reported in the 1730s, as we can see from the following, written in 1772 by the Reverend Thomas Comber, on the subject of an ovine disease that he referred to as rickets:

> The principal Symptom of the first Stage of this Distemper, is a Kind of Light-Headedness, which makes the affected Sheep appear much wilder than usual, when his Master or Shepherd, as well as a Stranger, approaches him. He bounces up suddenly from his Laire, and runs to a Distance, as though he were pursued by Dogs, &c. . . .
>
> In the second Stage of the Distemper, the principal Symptom of the Sheep is his rubbing himself against Trees, Posts, &c. with such Fury as to pull off his Wool and tear away his Flesh.
>
> The distressed Animal has now a violent Itching in his Skin . . . but it does not appear that there is ever any cutaneous Eruption. . . .

The third and last Stage of this dreadful Malady seems to be only the Progress of Dissolution, after an unfavourable Crisis. The poor Animal, as condemned by Nature, appears stupid, separates from the Flock, *walks irregularly,* (whence probably the Name of this Disease, *Rickets*) generally lies, and eats little. These Symptoms increase in Degree till Death, which follows a general Consumption. . . .

I do not find, Sir, that this Distemper is *infectious:* but alas! it is *hereditary,* and equally from Sire and Dam; and, like other hereditary Distempers, may lie latent one Generation . . . and then revives with all its former Fury. . . .

. . .

It is an incontrovertible Point, that whatever Sheep is once seized by this Distemper, never recovers; and it seems almost as incontrovertible, that whatever Sheep escapes it in his first Years, never takes it. . . .

. . .

This Distemper is generally said to be of about forty Years standing in *England;* and the Shepherds of this County pretend to trace it from the neighbouring County of *Lincoln* hither.[2]

And forty years before 1772 would take us to 1732. Independent confirmation of the presence of The Disease in Lincolnshire in the first half of the eighteenth century is found in a 1755 report addressed to the House of Commons by sheep farmers of that county. The report states that a disease here too called rickets—or shaking—had been affecting their flocks for ten years; that the disease was transmitted by rams; that it was often "in the blood" of their animals a year or two before it was detectable; and that once it had manifested itself, it could never be cured.[3] The farmers wanted measures to be taken against jobbers (speculators who had gained a monopoly on the trade in sheep) who mingled sick with healthy animals.

Following those early descriptions, the existence of The Disease was repeatedly reported through the late nineteenth century in Great Britain, Germany, and France. Oddly, it seems every so often to have been rediscovered as a new disease, which caused it to acquire numerous names. By the end of the nineteenth century, the English had settled on the name *scrapie,* and the French on *tremblante* ("the shakes").

The fact that the disease was seemingly forgotten between rediscoveries was due in large part, it appears, to its having been considered so shameful that farmers took great pains to conceal it. A single animal suffering from The Disease cast suspicion on the entire flock, considerably diminishing its value. For the farmer, it was both an economic disaster and a blemish on his honor. It must be said that the sight of a stricken animal was a poignant one, especially in the final stages of the disease. Here is a description written in 1937 by three prominent French veterinarians—Ivan Bertrand, Henri Carré, and Felix Charles Eugene Lucam—which is more detailed than Comber's account of a century and a half earlier but is clearly describing the same disease:

> When an animal is stricken, it scratches frantically, vigorously rubbing its tail, rump, lower back, and back against the walls or against its trough. Sometimes, it sits like a dog and energetically rubs the hind portions of its legs against the ground. Using its hind limbs, it scratches its head and the forward parts of its body. Using its teeth, it scratches the lower parts of its limbs. A continual victim of this generalized and persistent itching, the animal spends all its time trying to scratch itself, no matter how. . . .
>
> If the skin is examined at this stage of the disease, absolutely no lesions will be found: its smoothness, fineness, and coloring are intact, and it is absolutely impossible to connect this itching to a cutaneous cause.
>
> The sick [animals] seem bewildered and have a wild look in their eyes. Some suddenly begin to run, as if frightened, without cause. When they are alone and can be observed without their suspecting the presence of an observer, for example in the sheepfold, they are seen to be sometimes immobile, head high, ears alert, gaze fixed, as though they were hearing a distant noise. Then they suddenly jump and wildly make to flee an imaginary threat. During such flight, the

gait is most particular: the head is held very high, and the forelegs are flung far forward in order to cover ground. . . .

Their bleating is altered: indistinct, tremulous, and weak. In most cases, even the lightest touch, especially on the hindquarters, or the approach of a person or a dog will cause shuddering, quivering, or even intense and prolonged shaking. It is this localized or generalized muscular shaking that gives its name to the disease: *tremblante*. . . .

A second stage [of the disease] is characterized by more pronounced shaking of the head and the muscular system, by general weakening, by the appearance of secondary lesions from scratching, and above all by a new symptom: lack of coordination in movement. . . . Appetite, which has been steady until this point, begins to decrease; the animal loses weight and muscle tone decreases; this is the beginning of the cachexia [emaciation] that will continue to increase and that will become extreme in the final stage. . . . Because pruritis [itching] becomes more intense, the animal rubs, scratches, and bites itself to the point of damaging the skin. Owing to the constant rubbing, the wool becomes brittle and wispy and eventually is torn out over large areas. In those bare areas, the irritated skin reddens, thickens, wrinkles, and is covered with scabs. Scratches and open wounds appear, and there suppuration begins.

The animal is soon in an appalling state of emaciation and uncleanness, with remnants of shaggy fleece and bare skin covered with scabs.

Then lack of coordination in movement is seen: gradually the sick [animal's] gait becomes unsure and hesitant; it remains in the rear of the flock and follows it with difficulty; it stumbles with every step. Locomotive disorders are to be observed especially in the hind legs, which move with difficulty and are stiff. . . . If its gait is quickened, movement is confused, with the forelegs trotting and the hind legs galloping. . . . Falls are frequent. . . .

In the third stage, all the symptoms previously mentioned grow worse. . . . [The animal] staggers as though drunk, and prefers to remain lying in a corner. Sometimes, when it is forced to stand unaided, it will remain immobile, its limbs spread, its head lowered, and its body gently swaying as though it were trying to keep its balance on a moving platform; then, zigzagging with difficulty, it will reach a corner where it will fall in a heap, often uttering a moan.

In the final stage, rising and moving about become impossible, and the sick [animal] is able only to crawl on its knees. Emaciation is extreme and appetite has completely disappeared. Often, fetid and exhausting diarrhea sets in. The animal ends by stretching full out on its side; occasionally, it moves its limbs—which indicates clearly that paralysis never occurs. . . . Body temperature decreases, and death occurs without death throes, with complete physical decline.

We note that no thermal reaction is present at any point in the disease. . . .

The time between the appearance of the first symptoms and death can be from six weeks to six months. On average, it is about three months.[4]

That description echoes the symptoms set out by many writers since the early eighteenth century. The relative importance attributed to the various symptoms is, however, not constant, and this is reflected in the variety of names for the disease. *Scrapie* comes from "to scrape," laying stress on the apparently unbearable itching to which affected animals are prey, which causes them to scratch wildly to the point of tearing out patches of wool. That is the symptom highlighted also in the old French term *prurigo lombaire,* lumbar prurigo ("prurigo" being any of several kinds of itchy skin eruptions). On the other hand, other names used in France—such as *maladie convulsive, maladie folle* (mad), *maladie nerveuse, maladie chancelante* (wobbling), *névralgie lombaire* (lumbar neuralgia), *trembleuse,* and, of course, the current term *tremblante*—focus on neurological symptoms: The nervous system is obviously affected. That aspect of the disease is reflected also in another, rather euphonious, term used in France, *vertige du mouton*—ovine vertigo or "sheep dizziness"—as well as in the most common German term, *Traberkrankheit. Traber* means "trot," and the name reflects the characteristic gait of sick animals.

Because of this diversity in the way the symptoms were perceived, only very belatedly was it realized that this was but one disease. It also makes it risky to identify scrapie among the ovine ailments described before the eighteenth century and even casts doubt on the true nature of diseases described as scrapie or *tremblante* in the eighteenth and nineteenth centuries. Indeed, some symptoms of scrapie, taken in isolation, could be confused with those of other diseases.

Such uncertainty is one reason why it is so hard to date scrapie's arrival in western Europe. Some suggest that it came with the importation of merino sheep from Spain with a view to producing high-quality wool. This took place in England early in the eighteenth century, but in France not until the end of that century, which would seem to correspond to the dates of the first descriptions of the disease in each country. Others question the involvement of these merinos, and consider that their importation merely occurred when great attention was beginning to be paid to sheep farming, which in turn led to detection of the disease. In any event, most people agree that scrapie would have existed in Germany and central Europe before merino sheep were introduced to those areas.

For the farmers and veterinarians of the day, the key problem was to find a response to the disease, which could claim between 5 and 10 percent of some flocks. The disease was always fatal, and no matter what treatment was tried, it failed. It was important first and foremost to understand the cause of the disease so as to be able to protect animals from it. Here, there was disagreement, to say the least. Some saw scrapie as an infectious disease, some thought it was hereditary, and others linked it to environmental factors, diet, or the conditions in which the animals were bred.

Among the proponents of the infectious nature of the disease was a German writer who, in 1759, suggested that the best solution for a sheep farmer who discovered one of his animals to be suffering from scrapie was to remove it immediately from the flock, slaughter it— and use the meat to feed the servants. He added that the sick animal should be isolated without delay because the disease was contagious and could cause grave damage in the remainder of the flock.

Others entirely rejected the notion of contagion—noting, for example, that in a given flock the offspring of some rams were affected while those of other rams were not. From this they concluded that the disease was hereditary.

Then there were those who believed in neither contagion nor heredity, including one Monsieur Lezius, whose opinion was described in 1827:

> From his very precise observations, Monsieur Lezius concluded that vertigo in ewes results from an evil practice followed at the time of mating, and that this disease particularly affects ewes sired by excessively ardent rams which, in their overexcited state, are prevented from adequately satisfying their reproductive instincts. Such rams, enabled to cover only one or two ewes a day, will have a great number of offspring afflicted with vertigo; those who cover several will have fewer. Finally, it is probable that those left alone, without rivals, in a sufficiently large flock will sire none at all.[5]

So, scrapie would seem to result from sexual frustration among rams.

The veterinarian Roche-Lubin, who practiced at Saint-Affrique in the Languedoc region of France, had a different view, of which he wrote in 1848:

> In our land, the causes of scrapie are excessive copulation by rams; the rough fighting in which they engage amongst themselves; the sustained use of feeds that arouse them; leaping; violent exertion; rapid running when being chased by dogs; loud thunder; bright sunshine in the first few days after shearing; and the frequent recurrence of heat among infertile [females].
>
> Furthermore, scrapie is sometimes observed following difficult births; following aborted pregnancies in the first stage of gestation; after recovery or during convalescence from certain intestinal inflammations; after excision of mammary tissue in cases of gangrenous mastitis. . . .

I have never seen a case of scrapie without the existence of one of those causes, which undoubtedly modify, to a greater or lesser extent, the situation by acting slowly or less slowly, but in stages, on the nervous system.[6]

Today, that analysis seems laughable. Perhaps most surprising is the lack of points of comparison—control groups. Indeed, what sheep has never been exposed to one or another of the many causes identified by Roche-Lubin—for example being chased by a dog or hearing a thunderclap? Yet not all sheep have scrapie. We can see from Roche-Lubin's findings how scientists of the day could draw conclusions on the basis not of properly conducted experimentation but of their own preconceived ideas and beliefs. The central role assigned to the sex life of rams—viewed as frustrated by some and as excessively lustful by others—inevitably reflected the writers' moral or religious beliefs.

In 1848, the very year that Roche-Lubin published his observations on scrapie, a paper was submitted to the Académie des Sciences, titled "On the Possible Relationship between Crystalline Form and Chemical Composition, and on the Cause of Rotational Polarization."[7] This was the first publication of a man whose work would revolutionize the approach taken in both human and veterinary medicine: Louis Pasteur. He never so much as mentioned scrapie in any of his writings, but his work would provide the conceptual framework for study of that disease to this day. The work of Pasteur and his disciples, which we shall discuss in the next chapter, finally made possible the truly scientific investigation of the causes of scrapie.

2

MOLECULES AND MICROBES

PASTEUR'S EARLY WORK related to what we now call physical chemistry. Before Pasteur, chemists had established that substances—solids, liquids, and gases alike—were in general composed of molecules, which were in turn composed of atoms. A so-called pure substance contains molecules of a single kind, each containing a particular number of atoms linked by very specific bonds. As Antoine Lavoisier (1743–1794) demonstrated in his work during 1783–1785, for example, a molecule of water consists of one oxygen atom and two hydrogen atoms. However, in studying a compound somewhat more complex than water—tartrate, which is deposited in fermentation vats—Pasteur concluded that a molecule is not always completely defined by the atoms it contains and the bonds that link them; their spatial arrangement within the molecule is also a factor. Thus, tartrate could exist in two forms, corresponding to two kinds of molecules each containing the same atoms; each of the two kinds of molecules would be asymmetrical in itself but symmetrical vis-à-vis the other, like our left and right hands.

The asymmetry of these molecules gave rise to a specific, easily measured optical property of tartrate solutions: rotation of the plane of polarization of light. By measuring this effect, Pasteur observed that many compounds were, like tartrate, composed of asymmetrical molecules, and that all of these were of either plant or animal origin. In retrospect that observation makes sense, because the asymmetry that Pasteur observed was a result of very specific properties of the carbon atom, which is present in all organic molecules—that is, molecules produced by living things. For Pasteur, asymmetry became the hallmark of life, and it was this belief that would draw him to the study of fermentation and then of infectious disease.

Fermentation had been known since antiquity and was used to make bread, wine, and many other foods and beverages. But at the time Pasteur began to study the subject, views of fermentation were greatly confused. Since the invention of the microscope in the late seventeenth century it had been observed that microscopic creatures—known variously as animalculi, globules, mycodermic vegetables, or yeasts, among other names—were present in media undergoing fermentation. But no one understood the meaning of that phenomenon. Pasteur observed that compounds that caused the plane of polarization to rotate would appear or disappear during the fermentation process. In his view these compounds must be either produced or consumed by living things, leading him to conclude that fermentation resulted from the development and multiplication of microorganisms present in such media. He showed that these microorganisms could be cultured in defined media, and that it was possible at any time to trigger fermentation by inoculating a medium with such a culture. Moreover, he demonstrated that each type of fermentation corresponded to a specific microorganism. The organism that caused the sugar in grape juice to turn into the alcohol in wine, for example, was not the same as the one that turned wine into vinegar.

Pasteur needed to understand where these microorganisms came from. Many people thought that these microorganisms, which were seen as rudimentary life-forms, simply appeared spontaneously in environments favorable to their development. This was the theory of spontaneous generation. (Let us not forget: It had been only recently that mice were thought to appear out of nowhere in a laundry basket left lying around in the loft.) Through extremely rigorous experimentation, Pasteur demonstrated that the appearance of microorganisms in a previously sterilized medium could always be explained by introduction of "germs" from outside that medium. Most often, the germs—the microorganisms—were borne on fine dust particles suspended in the air. Pasteur thus undermined the experimental basis of the theory of spontaneous generation.

The focus of investigations into spontaneous generation would soon turn from fermentation to contagious disease. Before Pasteur, many scientists had noted the similarities between the two. Pasteur himself stated this upon demonstrating that there could be no fermentation in grape juice in which the introduction of environmental yeasts had been prevented: "Might it not be permitted to believe, by analogy, that the day will come when easily utilized preventive measures will end these scourges which abruptly afflict and terrify people, such as yellow fever and bubonic plague?"[1]

From that point, Pasteur set out to show that, like fermentation, contagious diseases were caused by microorganisms, and that each sickness had its own germ. His work, and that of Robert Koch (1843–1910) in Germany, laid down the rules for establishing a causal link between a germ and a disease: verifying the presence of the germ in all cases where the pathology existed; isolating the germ in pure culture; and reproducing the disease solely by using that germ. Those rules were initially applied to the study of a veterinary disease, anthrax.

Anthrax—whose name comes from the Greek for "coal" and refers to the very dark color of the sick animals' blood—caused serious damage among populations of sheep and cattle. The work of Pasteur and Koch proved that it was caused by a microorganism, in this case a bacterium known today as *Bacillus anthracis*. The mechanism of transmission posed problems for Pasteur, however, which deserve our attention because they are not unrelated to very timely issues related to the mad cow crisis. We shall return to these in Chapter 3.

Through his work on anthrax and a number of other veterinary diseases, Pasteur showed that it was possible to prepare vaccines against those diseases by "attenuating," or weakening, the germs that caused them. He grew anthrax bacteria in controlled conditions and thus obtained a germ that, when an animal was inoculated with it (when the germ was introduced into the animal's system) would make that animal resistant to the virulent form of the bacillus. The development of an anthrax vaccine had a considerable impact, but doubts remained about Pasteur's theories. In a bid to convince the skeptics, Pasteur turned his attention to rabies, which affected humans as well as animals. Its symptoms in human patients gave rise to a certain fascination, as expressed by one of Pasteur's biographers: "Rabies stirs our imagination. It evokes images of legend and of frenzied patients terrorizing all those around them, tied up and screaming—or suffocated between two mattresses."[2] For Pasteur, conquering rabies would prove his theories once and for all.

But first he had to identify the germ that caused it (and his procedures would provide a model for the study of scrapie some years later). Pasteur began by noting that rabies was transmitted by biting, so perhaps the germ would be found in saliva. Under his microscope, Pasteur examined the saliva of rabid dogs. He saw microbes, to be sure, but he saw the same ones in the saliva of healthy dogs, so that was not the answer. Well, if rabies affected the nervous system, perhaps the microbe might be found there. Here again, the microscope could not detect a rabies germ. But healthy animals would develop rabies if their brains were injected with saliva from rabid dogs or with ground brain tissue from dogs that had died of rabies. And their saliva or brain tissue could, in turn, cause rabies in other animals, and so on. This suggested that the invisible microbe multiplied whenever it passed into another animal; had it been an inanimate poison, it would quickly be diluted through successive generations and would become ineffective. Pasteur was able, in a way, to grow this microbe in the nervous system of living animals, but he was unable to do the same in any culture medium, so the mysterious rabies germ remained elusive. That did not prevent Pasteur, however, from making a vaccine from the spinal cords of rabbits inoculated with the disease. This was his final great triumph, which won him universal glory and the title of benefactor of mankind. But that is another story.

What, then, was this microbe that was invisible to the microscope but that caused rabies? The key was not found until the early twentieth century. At the turn of that century, it began to be understood that some microbes were so small that they could pass through filters that would hold back the usual kind of germs, such as those of anthrax, plague, and cholera. These ultramicroscopic microbes were called viruses, a name that had previously been very loosely used to cover anything that transmitted diseases but whose nature was not known. Viruses could not be seen with optical microscopes, but the invention of the electron microscope in 1933 made it possible to observe them. Only then could we finally "discover" the rabies virus that Pasteur had used—even though he was never able to see it himself—to prepare his vaccine half a century earlier.

3

MAD DOGS AND EARTHWORMS

PASTEUR AND KOCH had established that anthrax was caused by a bacterium, but how was it transmitted within a flock? That would also be the question a century later for scrapie and BSE.

In some flocks, anthrax is endemic; from time to time, an animal will succumb to it. In other flocks, it affects a large number of animals within a short space of time, thus becoming epidemic. Furthermore, it can flare up periodically or remain dormant for many years. Koch reported two important facts that advanced our understanding of these phenomena: First, he observed that in conditions not conducive to multiplication, the bacillus can produce spores—a very resilient form of the bacterium—enabling it to slumber for long periods and to begin again to multiply when conditions again become favorable. Second, he was able to cause the disease in animals by mixing the bacillus or its spores into their food. Contamination could thus occur through feeding, and possibly even with spores that were several years old.

Still unknown was exactly how this occurred in nature, and especially the significance of fields that were said to be "cursed with anthrax," in which flocks could not graze without a good number of animals contracting the disease. A possible solution came to Pasteur during a walk:

The harvest had been brought in, and nothing was left but stubble. Pasteur's attention was drawn to one area of the field by the different color of the earth. The landowner explained that sheep that had died of anthrax had been buried there the previous year. Pasteur, who always liked to examine things at close quarters, noted on the ground a great number of worm castings. This gave him the idea that, during their unceasing travels from underground to the surface, the worms were bringing to the surface the humus-rich earth that surrounded the corpses and, with it, the anthrax spores it contained. . . . Pasteur never stopped with a mere idea: He immediately proceeded to experimentation. This bore out his predictions. The earth contained in one of the worms, when guinea pigs were inoculated with it, gave [the guinea pigs] anthrax.[1]

The spores that the worms brought to the surface contaminated the plants on which the animals grazed and could give them anthrax, especially if their diet included straw, stubble, thistles, or awns from heads of grain, which could cause small mouth lesions that enabled the germs to enter through the mucous membranes.

The transmission of anthrax thus seemed to have been explained, which was important from the theoretical standpoint, because many still doubted Pasteur's germ theory. They were known as "spontaneists" because, although they did not deny that "morbid viruses" could spread within a population, they believed that these viruses initially appeared spontaneously in the body owing to abnormal physical, physiological, or dietary conditions.

And sometimes, the facts made it seem as though they were right: In the very case of anthrax, the veterinarian Edmond Nocard, a disciple of Pasteur, in 1881 reported several cases that without his insight would certainly have been seen as examples of spontaneous generation. I shall cite just one of these; it is not without resonance in the matter before us today: Nocard wrote of "the imprudence of a new farmer, full of zeal and initiative and wishing for high yields, who (unheard of in that area) during his first year bought a large quantity of artificial fertilizer. His wheat was superb, but the following year, no sooner had he put his flocks to pasture on the fertilized ground than anthrax appeared, and that very year he lost nearly a quarter of his stock; subsequently the disease continued to wreak havoc."[2]

That fertilizer had been manufactured by a large company that processed animal corpses from miles around and that paid little attention when it turned them into fertilizer. From this, Nocard drew conclusions about the probable role in the spread of anthrax of the use of artificial fertilizers that had been badly prepared from infected animal wastes. Does this not eerily prefigure the role played by animal-based feeds in the spread of BSE a century later?

The 1870s saw contention between Pasteur—the "charlatan chemist" and leader of the "contagionists"—and the spontaneists, who included many of the physicians and veterinarians of the day. Prominent among them was Henri Bouley (1814–1885), director of the Ecole Vétérinaire d'Alfort, outside Paris, who repeatedly defended the idea of spontaneous generation of the "morbid viruses" responsible for many diseases in animals. He began to have doubts in 1874, but here is what Bouley wrote in that year in connection with rabies: "Can rabies in dogs develop spontaneously and, if so, is the principal—or sole—condition for this to be found in genital ardor constantly stimulated by the discharges of a bitch in heat but never assuaged?"[3]

Although he conceded problems with that theory, Bouley by no means rejected it. In accordance with the principle that an ounce of prevention is worth a pound of cure, he felt that necessary prophylactic measures should be taken: "Could one not, for example, prevent bitches in heat from wandering the city streets and from igniting, with their discharges, the fires that rage in their pursuers? If one among these is more easily aroused and more ardent than the others, is it not possible that the passion that consumes him—the madness of love—could be transformed into true madness?" And this: "It would seem to us a wise precautionary measure, for example, not to leave male dogs in proximity to their females when the latter are in heat if a more intimate liaison is forbidden them for one reason or another."

Assigning sex a major role in the onset of diseases of the nervous system—which we have already seen in connection with scrapie—is part of a trend that prefigures the work of Sigmund Freud (1856–1939). But for rabies, it was off the mark, as François Tabourin, professor of physics and chemistry at the Ecole Vétérinaire de Lyon, tried to convince Bouley:

> What hasn't been put forward to explain the supposed spontaneous onset of rabies? In turn, heat and cold, dryness and humidity, winter and summer, spring and autumn, the sun and the moon—not to mention the stars—have all been proposed. But statistics, which are without imagination—have demolished the supposed causes of supposed spontaneous rabies. What other causes have been dreamed up to explain the onset of rabies without inoculation? Oh, I almost forgot the most original: muzzling. Yes, Mr. Editor, . . . muzzling induces the onset of rabies! . . . But the great hobbyhorse of the advocates of the spontaneous nature of rabies is reproductive ardor in dogs that is not satisfied after having been strongly aroused.[4]

Tabourin then proceeded to refute the theory. He states that statistics showed that only one or two rabid dogs per thousand were claimed to suffer from spontaneous rabies, having contracted the disease without a known source of contamination. He expresses astonishment that a theory could be founded on so small a number of cases: Rabies is spontaneous in one case in a thousand, while resulting from infection by another rabid dog in all other cases? And what about those exceptions? Had there really been no contact with other dogs? He gives an example of a "bitch whose masters had never let it out of their sight" and which had therefore been said to die of spontaneous rabies, "but which, on autopsy, was found to be pregnant." Then there was another dog, which, "on the testimony of a servant charged with looking after it, had for ten years been thought to have died of spontaneous rabies, but which had to be stripped of that honor after a new formal statement by the servant, who—no longer needing to conceal the truth—affirmed that the dog had been bitten."

At the end of his long letter laying into spontaneist theories, from which the foregoing quotations are drawn, Tabourin offers a series of proposals on the formulation of new health legislation aimed at limiting the effects of contagious diseases among domestic animals. The provisions he wanted to see in the new law have strange resonance today:

> Article 1: All those who possess animals affected by a contagious disease shall declare this to local authorities as soon as possible. All those who neglect to meet this obligation shall lose their right to compensation as specified below.
>
> Article 2: All animals affected by a contagious disease, as well as those which live together with them, shall be sacrificed should the public interest require it, once their condition has been duly certified. One or several veterinarians, depending on the seriousness of the cases, shall be appointed to visit the animals affected by a contagious disease and to certify their condition.
>
> Article 3: Compensation equal to two thirds of the monetary value of the animals shall be granted to the owner of animals sacrificed in the public interest. That value shall be determined by a committee composed of a veterinarian appointed by the authorities and by two owners selected by the mayor of the district in which the appraisal is taking place.
>
> Article 4: The places inhabited by sick or suspect animals shall be methodically disinfected under the supervision of a professional. Those places may not be occupied by other animals for between fifteen and thirty days, depending on the degree of infectiousness of the contagious disease.

Four years later, in 1878, Tabourin resumed his attack on the spontaneist theories that Pasteur had still failed to eradicate among veterinarians, and concluded his article with these words: "Moreover, in our view, contagion and spontaneity are two irreconcilable things, and because such diseases have at their basis a kind of germ that can somehow be sown by the inoculation of ground propitious to its development, our mind refuses to accept the birth of a specific disease without the original involvement of its necessary seed—any more than it would accept the sprouting of a blade of wheat without the assistance of a seed. Thus, in our view, any disease that can be sown cannot develop spontaneously."[5] Here Tabourin set out what remains the credo of doctors and veterinarians to this day. Yet today, scrapie could call it into question once again.

But let us return to the matter at hand. Those on the trail of The Disease would then take a more scientific turn. Pasteur had shown the path; they only needed to follow it.

4

SCRAPIE UNDER THE MICROSCOPE

IN 1898, THREE YEARS AFTER Pasteur's death, Professor Charles Besnoit of the École Vétérinaire d'Alfort learned that for several years an unknown disease had been ravaging flocks in the Tarn, in southwest France. Mortality had reached 15 to 20 percent, devastating for a region where sheep were important not only for wool and meat, but also for milk used in cheese production. Besnoit soon realized that this "unknown" disease was scrapie, and he decided to study it—becoming the first to do so in a genuinely scientific way. His principal achievement would be to resolve a question that had baffled all his predecessors: the nature of the organic lesions that were responsible for the symptoms.

Given the obvious symptoms described in previous chapters, one might have expected to observe visible changes in certain tissues. And as we have seen, skin lesions were unsuccessfully sought as an explanation for the severe itching that plagued sick animals. To understand the behavioral changes, scientists had examined various parts of the nervous system as well as other organs—unsuccessfully, as indicated by these conclusions drawn in 1830 by Jean Girard, director of the École Vétérinaire d'Alfort: "We are compelled to state that we observed no organic change that could confirm the presumed locus of the disorder. The spinal cord, the lumbar nerves, and the rachidian [spinal] girdle nearly always appeared to us to be entirely sound."[1]

When Besnoit began his work, he initially came to the same conclusion: "Upon autopsy, no macroscopic changes were noted. The brain, spinal cord, nerves and muscles appeared healthy."[2] He found "no macroscopic changes"—that is, none that were visible to the naked eye. But Besnoit looked further.

Since 1830, laboratory scientists had come to rely on the microscope. For Pasteur, it was his chief tool, and with it he revealed a whole new world, the world of microscopic organisms. But others were using it to examine fragments of living tissues, both animal and vegetable. They had observed that all these tissues were composed of constituents that came to be called cells. The cells of various tissues differed in appearance, shape, and size, but they had a number of common characteristics. They were enclosed in a membrane and contained a nucleus and many smaller components; all of those floated in an intracellular medium called cytoplasm.

So, unlike his predecessors, Besnoit did not limit himself to the naked eye in examining the organs of animals that had died of scrapie. He used his microscope, recording observations like this: "Upon microscopic examination, on the other hand, very obvious nervous system lesions were seen; these were located in the spinal cord and in the peripheral nerves." Among those lesions, Besnoit described bubble-like "vacuoles" in some nerve cells in the spinal cord. These were to become the signature of scrapie; they made it possible, on autopsy, to distinguish between animals with scrapie and those affected by other diseases.

Obviously, Besnoit did not stop there. Like his predecessors, he tried to learn what caused the disease. He first looked for bacteria in diseased tissues, both by direct microscopic examination and by inoculating a variety of culture media with nervous system tissue or with blood. The results were invariably negative. Then, he considered in turn the various other hypotheses. He felt that the breed of sheep could have an influence in terms of predisposition, but that it was not decisive. Although it had been suggested by many farmers, the possibility that the disease was transmitted by breeding rams, like a venereal disease, did not seem to stand up to analysis. The impact of diet, which he had considered previously, remained to be proven. He devoted particular attention to the question of whether the disease was infectious, but despite careful study he was unable to confirm that hypothesis. All attempts failed to reproduce the disease by inoculating a healthy animal with brain or medullar tissue or with blood from a sick animal. One healthy ewe was given nearly two liters of blood from a sick ewe in the terminal stages of the disease, but showed no symptoms of scrapie after nine months.

He also housed several sick ewes with two healthy ones for a period of six months, yet no sign of the disease could be detected in the healthy ewes. Besnoit concluded that it was "impossible at present to affirm that scrapie is of a microbial and infectious nature."[3] But the debate was not over; now it was the turn of the British to carry it forward.

One Scottish veterinarian, Sir John MacFadyean, reported on epidemiological observations that convinced him the disease was contagious. In a 1918 article, he recounted the appearance of scrapie in the flocks of two farmers, whom he referred to as Mr. A and Mr. B.

In his first eleven years of farming, according to MacFadyean's narrative, Mr. A had never had a case of scrapie among his flock, and he believed there had been none on neighboring farms. In autumn of 1907, Mr. A bought 140 young ewes at auction in a nearby town. They had been raised by one Mr. X; it was subsequently determined that scrapie existed in his flock. These young ewes were mated in November 1908, along with thirty other young ewes of the same age, bred by Mr. A on his own farm. In February 1909, the first symptoms of scrapie appeared in the flock, and in the course of the next six or seven months some thirty ewes died of that disease. They were all among the 140 ewes that Mr. A had bought in 1907.

Mr. A was determined to rid his flock of the disease, and he sold all the surviving ewes from that group for meat, along with their lambs. By autumn 1909 no ewes raised by Mr. X, or any of their offspring, remained on Mr. A's farm. At first, this seemed to have worked, for Mr. A experienced no new cases of scrapie for the next eighteen months. Unfortunately, this was nothing more than a respite: Two cases appeared in April 1911, to be followed by several more.

Very nearly the same happened to Mr. B, who had also bought ewes raised by Mr. X.

The foregoing facts were a powerful argument that scrapie was contagious. But if that were true, the incubation period had to be exceptionally long—at least eighteen months, which was the length of time both between the purchase of the ewes and the appearance of the first symptoms, and between the elimination of all ewes from the presumably infected group and the appearance of the first symptoms among the ewes raised by Mr. A himself. This eighteen-month-plus incubation period was consistent with the long-known fact that the disease affected only animals two years old or older.

Although he was convinced that scrapie was contagious, MacFadyean was no more successful than Besnoit in transmitting the disease by inoculating blood, brain or spinal tissue, or other substances from sick animals. The inability to isolate infectious samples made it pointless to search for a causative microbe. This, of course, did not stop the flow of speculation, as we can see from the less than agreeable exchanges between MacFadyean and another titled Scot who specialized in scrapie, Sir John Poole MacGowan. MacGowan thought that scrapie might result from a massive muscular infection by sarcosporidia parasites, but MacFadyean believed not a word of it and didn't hesitate to say so.

While our two Scots were quarreling about the nature of the scrapie microbe, the First World War was raging in Europe. Two centuries had passed since the Age of Enlightenment, when The Disease had been observed in English sheep. What had we learned about The Disease in those two hundred years?

First and foremost, we had learned to recognize it, at least in sheep and in spite of the varied disguises it could assume. The set of symptoms— associating severe itching with a variety of neurological symptoms—

provided what doctors call an overall clinical picture. Thanks to Besnoit's work, the diagnosis could be confirmed upon autopsy by the presence of characteristic lesions in certain nervous system cells, especially those of the spinal cord. On the other hand, the mystery remained virtually complete with respect to the causes of the disease—its etiology. Many facts suggested that it was contagious, but nearly as many ran counter to that hypothesis. Most attempts to intentionally infect healthy animals by penning them together with sick animals had failed, along with attempts to transmit scrapie using fluids or tissues from sick animals. Plus, it had proven impossible to identify any infectious agent whatsoever in affected tissues, either by microscopic examination or by culturing. Finally, if this was a contagious disease, its incubation period—eighteen months to two years at a minimum—was far longer than those observed in traditional contagious diseases, which ranged from a few days to a few weeks. For all those reasons, there was room for doubt about the contagious nature of scrapie.

In particular, it was impossible to rule out the possibility that the disease was at least partly genetic in origin, especially since many farmers had observed breed to have an influence. Some farmers felt that scrapie might be simultaneously hereditary and contagious. But at that time all discussions of heredity were characterized by great confusion: It was only the dawn of the science of genetics.

The father of genetics, Gregor Mendel (1822–1884), was a contemporary of Pasteur—they were both born in the same year—but his work remained unknown until the turn of the twentieth century. It had long been observed that living things gave birth to creatures similar to themselves. But until Mendel, the underlying mechanism of the stability of species was completely unknown. No one understood how traits were passed from generation to generation. Mendel's genius lay in his concentrating on a small number of easily recognized traits, crossing parents with differing traits and studying their offspring. He chose to experiment on peas, and looked at traits such as the color and shape of their seeds. By analyzing the distribution of these traits among the first-generation and second-generation offspring, he concluded that inheritance of the traits was carried by physical elements that would later (in 1909) be called genes. They existed in pairs in every cell in an organism, except in reproductive cells—spermatozoids and ova—where they existed singly.

Early in the twentieth century, a parallel was established between genes—which then remained hypothetical things whose existence had only been postulated by Mendel—and tiny structures that had been discovered in the nuclei of cells: chromosomes. Before a cell divides, all the chromosomes in its nucleus split, then the duplicate chromosomes separate. One "twin" will be found in each of the resulting daughter cells. While nothing was then known about the mechanisms of duplication, separation, and segregation of chromosomes in daughter cells, it was tempting to predict that the chromosomes contained genes. If genes were indeed passed from parent to offspring in order to transmit traits, they had also to be transmitted from parent cells to their descendants so that the traits transmitted to the fertilized ovum would be found in all cells to which it gave rise. This led to the chromosome theory of heredity, which likened chromosomes to strings of pearls, with each pearl being a gene responsible for a specific trait in the individual organism.

Genetics was to play an important part in tracking The Disease. But for the time being, it was still an emerging science, unknown to the lay public and not yet broadly understood even among scientists. But veterinarians could not fail to be highly interested in the new science, for it cast new light on every aspect of their profession. And in 1913, it was a perceptive veterinarian, Sir Stewart Stockman, who responded to claims that scrapie could be simultaneously hereditary and contagious: "I may point out . . . that no disease is known which is both hereditary and contagious, although the mistake is not unnatural in a lay mind, which does not always distinguish the difference between hereditary transmission and congenital infection."[4]

Stockman might not put it exactly that way today, but he was absolutely correct in the context of the day. A disease could be hereditary—

resulting from a genetic flaw—or contagious—resulting from a microbe—but not both at the same time. He was also right to think that the lay public did not necessarily distinguish between true hereditary transmission and congenital disease. Remember that many still believed tuberculosis could be hereditary, while the fact is that children were infected by their parents after birth.

With the Great War raging, let us for the moment abandon our French and British veterinarians to their work, and consider what was taking place on the opposing side, in Prussian Germany. There too, The Disease was lurking—but this time it was not only sheep that were involved.

5

CREUTZFELDT, JAKOB, AND OTHERS

THE POSTHUMOUS LEGACY of Hans Gerhard Creutzfeldt is a strange one. He died in 1964 virtually unknown, but two decades later his name was to be world famous—although some today question whether his renown is really justified. In any event, that fame is due to an obscure paper published in 1920 in a German neurological-psychological journal.

Creutzfeldt was an exceptional human being: an original, independent thinker, and a man whose brusque manner concealed great kindness and profound humility. "Knowledge makes you arrogant, but education makes you humble." Those words, spoken during his speech at the opening of the new University of Kiel, say a great deal about his personality. During the Second World War, Creutzfeldt did what he could to oppose the Nazi regime; his clinic at Kiel provided sanctuary for many who refused to submit to Hitler's racial laws. After the war, he exposed the particularly revolting activities of one of his colleagues, Werner Heyde, who had played a key role in carrying out the Nazis' euthanasia program targeting those mentally ill patients who were considered incurable. Under the alias of Fritz Sawade, Heyde had found work as a psychiatrist at Kiel and was often summoned to testify in court as an expert witness before his wartime activities were brought to light. (Heyde committed suicide in 1964, five days before he was to have been put on trial.)

Born in 1885, Creutzfeldt earned his doctorate in medicine at Kiel, then indulged his taste for adventure by serving as a naval physician in the Pacific. Upon his return he trained in neuropathology at Breslau (now Wrocław, Poland), Munich, and Berlin. It was at Breslau, in the department chaired by the celebrated Alois Alzheimer, that he examined a young woman referred to as Berta E. He reported his observations in the paper that would bring him posthumous glory: "On a Strange Focal Disease of the Central Nervous System."[1]

Berta E came to the University of Breslau clinic on June 20, 1913, aged twenty-three. She had previously experienced difficulty in walking normally and also manifested marked behavioral changes. She refused to eat or wash; she neglected herself. She had fallen when getting out of bed, but without losing consciousness. On another occasion, she suddenly shouted that she had caused the death of a nun in the convent she lived in. She thought herself demonically possessed. The night before her arrival at the clinic, she had been very agitated; she spoke a great deal, laughed, and sang. Soon after being hospitalized, it became impossible for her to walk unaided. She experienced tremors of the facial muscles and sudden, uncontrollable arm movements. She spoke incoherently and manifested disorientation both in space and in time. Her ability to understand slowed. She would make sudden grimaces and would be seized with causeless, mechanical fits of laughter. She was simultaneously apathetic and overly excitable. Beginning in mid-July, her condition rapidly deteriorated. Paralysis progressed and she no longer recognized those around her. In early August she experienced epileptic-type seizures. Her gaze became fixed and expressionless. She died on August 11. Upon autopsy, unusual brain lesions were observed. They appeared to be loci of degeneration, with neuron death throughout nearly all the gray matter. Creutzfeldt saw this as a previously undescribed neurological disease.

After his 1920 paper, Creutzfeldt published nothing more on this subject, and there was every likelihood that the article would molder on the shelf. It was salvaged by the work of another German neurologist, working at Hamburg: Alfons Maria Jakob.

Jakob was born in 1884, the son of a shopkeeper. He studied medicine at Munich, Berlin, and Strasbourg (then under German rule). In 1909 he earned his doctorate in medicine, specialized in neuropsychiatry, and, like Creutzfeldt, benefited from Alois Alzheimer's wisdom. In 1911 he moved to Hamburg, where he was to spend the rest of his career (apart from his service in the German army during the First World War). He was appointed professor of neurology in 1924, and was a respected teacher and an internationally known research scientist. This was not because of his study of what was soon to be called Creutzfeldt-Jakob disease, but because of his work on other neurological diseases such as neurosyphilis, multiple sclerosis, and Friedreich's ataxia.

In three articles published in 1921 and 1923, Jakob described the cases of two men and three women ranging from thirty-four to fifty-one years of age, all of them affected by a gradual impairment of motor function, speech, and emotions, with personality changes and loss of memory. They all ended up unable to move about, to stand, or to talk. Their intellectual faculties were profoundly impaired—they experienced dementia—and, after becoming bedridden, they died within a few weeks to a year after the onset of the most serious symptoms. The overall title of Jakob's three articles was "On a Strange Disease of the Central Nervous System, with Unusual Anatomical Observations."[2]

A particularly poignant case described in those articles was that of a forty-two-year-old soldier stationed in Romania on the shores of the Black Sea. He regularly wrote home to his wife in Germany. In May 1918 he began to complain of various problems such as dizzy spells and weakness. Starting in late August of that year, his wife observed that his handwriting was changing. The progress of his disease can be followed by simply reading his letters, first of all in terms of their content: He described his symptoms, and, moreover, the text reflects his gradual loss of reason. And second, we see it in his handwriting, which quickly deteriorated and by October had become illegible. The soldier was sent home to Germany in early December. He was hospitalized in Munich, where he was placed under Jakob's care. His condition gradually worsened and he died toward the beginning of March 1919.

Jakob himself took note of the similarity between the cases he was describing and that of Berta E, described by Creutzfeldt. So it was no surprise that yet another German neurologist, Walther Spielmeyer, in whose department Creutzfeldt had completed his work on the Berta E case, made reference to "Creutzfeldt-Jakob disease":

> The singular disease—with foci of degeneration, affecting the cerebral cortex and involving spasms, hyperalgesia [increased sensitivity to pain stimuli], and mental problems—described by Creutzfeldt has not remained unique. A. Jakob, through painstakingly analyzed data, has found many cases to add to it. We can thus hope that Creutzfeldt-Jakob disease (Jakob's spastic pseudosclerosis) will come to be thoroughly defined, in anatomical and clinical terms.[3]

It was years before Spielmeyer's hope would be fulfilled. For example, a 1998 letter to the journal *Nature* argued that the disease described by Creutzfeldt was not in fact Creutzfeldt-Jakob disease. Creutzfeldt himself had acknowledged after the Second World War that "his case did not bear any resemblance to the cases described by Jakob."[4] Furthermore, according to that letter, it appears that only two of the five cases described by Jakob were true cases of Creutzfeldt-Jakob disease as we understand it today. This shows the great confusion that reigned for many years with respect to identification of the illness, similar to the preceding two centuries' worth of confusion about scrapie.

So the disease continued to assume its disguises—for this was indeed The Disease, even though neither Creutzfeldt nor Jakob knew it.

For four decades there was hardly any follow-up to Creutzfeldt's and Jakob's work. At most, we can cite descriptions of four additional cases, in 1924 and 1930, by colleagues of Jakob, who died an early death in 1931. Among these four cases is that of a man named P. Backer, whose sister appears to have suffered from the same disease. The possibility that the disease was genetic was raised at the time, and this was considered again in 1950 by the team that succeeded Jakob's at the University of Hamburg. It was known that not only the sister but several other members of Backer's family had died of the same illness. Under the theory that the disease was genetic in origin, the mutation responsible for it would have to be dominant. But to understand this concept, we must return for a moment to Mendel.

Mendel stated that the various traits of a plant or animal were determined by elements now known as genes, which are present, in duplicate, in each of an organism's cells, apart from in the reproductive cells, which contain only a single example of each gene. When a female reproductive cell is fertilized by its male counterpart, the resulting zygote— the fertilized egg—again contains two examples of each gene, one from the father and one from the mother. In principle, they both contribute to the expression of a trait. For example, in determining the color of the seed of a pea plant, the two examples of a gene would contribute to the formation of the pigment. But one of those genes might carry a mutation that makes it incapable of participating in the formation of that pigment. When the other gene by itself makes possible the formation of more or less normal quantities of pigment, that gene—called an allele— is referred to as "dominant." This means that, when it is present, the mutation in the other gene goes unnoticed. The other gene is referred to as "recessive."

So long as a dominant allele is present in a given organism, the presence of a recessive allele (carrying a recessive mutation) cannot be detected. But if both examples of the gene carry the mutation, its effects will be felt. For instance, consider a scenario in which two apparently normal organisms are bred, but each carries a normal dominant and an altered recessive example of a given gene. In such organisms half the reproductive cells will carry the dominant allele and half the recessive one.

When these are crossed, whenever a male reproductive cell containing the recessive allele fertilizes a female reproductive cell containing the same recessive allele, the resulting zygote will contain two examples of the recessive allele. To return to the example of a gene responsible for pigment formation in peas, the fertilized ovum will yield an organism without pigment and hence without color. It is a simple matter to calculate that in such crossing one-quarter of offspring will inherit two recessive alleles and will thus be colorless, while all the others will have color, either because they inherited two dominant alleles (one-quarter of them will do so), or because, like their parents, they have one dominant and one recessive allele (which accounts for the remaining half). In fact, Mendel's thinking followed the same sequence in reverse, by analyzing the various traits seen in the offspring of such crossings, and thus postulated the existence of genes.

To get back to the Backer family, if the disease that affected its members was genetic in origin, that meant it had to result from a mutation in a gene. If the mutation was dominant, its presence in just one of the two examples of that gene would suffice to bring on the disease. So any child of an affected individual had a fifty-fifty chance of inheriting the altered allele and thus a fifty-fifty chance of contracting the disease. Dominance would suggest that the alteration of the gene involved not a loss of function but the appearance of a new function that caused the disease.

While Creutzfeldt's and Jakob's articles were gathering dust on library shelves, other neurologists were making interesting observations in a Vienna psychiatric hospital. In 1928, Josef Gerstmann described the case of a twenty-five-year-old woman who had symptoms of what he took to be an unusual disorder of the cerebellum. He had observed a strange reflex in that patient: If she turned her head while holding her arms out in front of her, she would automatically cross her arms, and the arm opposite the side toward which her head was turned would always be on top. In 1936 Josef Gerstmann, Ernst Sträussler, and I. Mark Scheinker described this case in detail, along with cases involving seven other members of the same family. Some of the symptoms recalled those observed by Creutzfeldt and Jakob, but the authors made no mention of their work. But other symptoms were entirely different, as were the lesions observed in nervous system tissues. Among these, the authors described clusters of matter in the brain in the form of plaques that reminded them of similar formations seen in the brains of individuals who had died of Alzheimer's disease. They concluded that this was a new disease of the central nervous system, most likely hereditary in nature.

But it was nothing other than The Disease in a new guise.

It was in sheep that we began to close in on The Disease, which Creutzfeldt, Jakob, and the others had unknowingly detected in humans: The forty or fifty years during which physicians ignored the observations of Creutzfeldt and Jakob were not wasted by veterinarians. So, let us return to the France of 1936, where the chase continued despite the Great Depression and signs of a coming world war.

6

SCRAPIE IS INOCULABLE

ON DECEMBER 28, 1936, the French veterinarians Jean Cuillé and Paul-Louis Chelle presented to the Académie des Sciences a communication titled "Is the Disease Known as Scrapie Inoculable?" They reported having used a variety of techniques to inoculate nine sheep of both sexes with cerebral or medullar matter from a number of animals in the latter stages of scrapie. In the course of the nine months following the inoculation, seven died or were sacrificed—for reasons beyond the control of the experimenters—and showed no sign of scrapie. Despite those discouraging beginnings, Cuillé and Chelle continued their observation of the two surviving sheep, both ewes. Their patience was rewarded, because they were to succeed where all their predecessors had failed. Here is an excerpt from their communication:

> Ewe number 1, aged two and a half years, was from a disease-free flock from the Narbonne area [in southwestern France] and had been living in the School's sheepfold for a year and a half. On July 9, 1934, she was given an intraocular inoculation of 3cm³ of an emulsion of spinal cord tissue that had been finely crushed in a mortar, with a small quantity of sterile saline solution.

Her general condition remained satisfactory until the latter half of September 1935. At that time, i.e., fourteen and a half months after inoculation, the animal became restless and walked with its head raised and with a frightened appearance. Two weeks later, in early October, locomotive problems arose, accompanied by senile-looking tremors of the head and neck.

By October 15, the symptoms were entirely characteristic of scrapie: severe lack of coordination of the hindquarters; exaggerated flexion of the forelimbs (a stepping gait); at high speeds, dissociation of movement of the fore- and hindquarters (trotting gait in the fore-limbs, galloping gait in the rear limbs); exaggerated movements of the head and neck; grinding of teeth. . . .

Having reached the final stages of the disease, the animal was sacrificed on October 30, 1935, nearly sixteen months after inoculation and a month and a half after the first characteristic symptoms had been noted.[1]

Ewe number 2 was inoculated in the same way and showed the first symptoms of disease twenty-two months after inoculation, dying two months later (about two years after inoculation). The authors drew the following conclusions from their experiment: Scrapie is infectious and inoculable; the virus resides in the nervous system—the spinal cord and the brain; the incubation period is long—fourteen and twenty-two months in these two cases.

These conclusions were met with a degree of skepticism, especially by three other French veterinarians—Ivan Bertrand, Henri Carré, and Felix Charles Eugene Lucam—who had just failed to cause the disease to be transmitted. In a September 1937 article, in fact, they described a series of five experiments in which they inoculated the brains of healthy animals with nervous system tissues from sick animals. In no case did the inoculated animals develop scrapie, and the authors concluded that, in spite of the claims of Cuillé and Chelle, "scrapie is not transmissible experimentally in sheep . . . through inoculation of substances from sick animals."[2]

But upon closer examination the contradiction between the two sets of experimental results is not as great as it might seem. In four of the experiments described by Bertrand, Carré, and Lucam, the inoculated animals were followed for a maximum of only three months—far short of the fourteen or twenty-two months it took for the disease to appear in the ewes inoculated by Cuillé and Chelle. Only one case seemed to contradict their observations: One of the ewes inoculated by Bertrand, Carré, and Lucam was kept under observation for twenty-three months without manifesting the least symptom of scrapie. But even there, a longer observation period might well have yielded the appearance of initial symptoms.

Impatience was probably at the root of many of the failures experienced by Cuillé and Chelle's predecessors in trying to transmit the disease. But there could have been other reasons as well—as, for example, in the many failed attempts by Sir John MacFadyean. As we have seen, he was among the first to offer convincing arguments that scrapie was contagious. Moreover, his observations suggested that if scrapie was indeed contagious, the incubation period would have to be at least eighteen months. In the series of sixteen experiments he described in 1918, he therefore took the precaution of observing inoculated animals for two to two and a half years. His failure to transmit the disease seems in retrospect to be the result of his inoculations having used not crushed nervous system tissues but blood or other substances of far less infectious potency (or none at all). He would have avoided this mistake if he had paid more attention to Pasteur's work on rabies, as Cuillé and Chelle undoubtedly did. Once Pasteur realized that the rabies virus grew in the nervous system, he used crushed spinal cord and brain tissue to transmit the disease. Are we, then, to attribute MacFadyean's error to his having worked in English while Pasteur and Besnoit published their findings in French?

Be that as it may, Cuillé and Chelle needed to confirm their preliminary results, which they did in January 1938 when they reported successfully transmitting scrapie to a ram and two ewes with incubation periods of eleven, twelve, and nineteen and a half months, respectively. The two ewes were inoculated with an emulsion of spinal cord tissue from one of the ewes to which scrapie had been transmitted by inoculation during the first experiment, while the ram was inoculated with material from an animal that had spontaneously contracted the disease. Three other animals inoculated as part of the same experiment remained scrapie-free after twenty-six months of incubation. It would seem, then, that either infection did not occur systematically or it possibly required even longer incubation periods.

The results attained by Cuillé and Chelle could no longer reasonably be questioned; scrapie could be transmitted in an experimental context, and had to be viewed as an infectious disease. It remained, however, to identify the responsible agent: the scrapie "microbe." Like Pasteur when he studied rabies, Cuillé and Chelle were dealing with an invisible microbe. Was it one of the so-called filterable viruses that had recently been discovered? The two veterinarians believed the answer was yes, and they succeeded in transmitting scrapie to two lambs by inoculating them with a filtrate from crushed spinal cord tissue from a sick ewe.

In another important achievement, Cuillé and Chelle succeeded in transmitting scrapie to two goats, one male and one female. The incubation period was only a little longer than in sheep: twenty-five and twenty-six months, respectively. This was the first time that scrapie had been described in goats, and a few years later Chelle would describe an instance of natural transmission in a female goat raised among a flock of sheep in which the disease had been present for several years.

Although Cuillé and Chelle were able to transmit scrapie from sheep to goats, they were unable to transmit it to rabbits. Bertrand, Carré, and Lucam had tried without success to transmit the disease to rabbits, guinea pigs, and white mice. This failure was surely one of the main reasons for the deafening silence that followed the major discoveries we have just described. The time was ripe for an attempt to purify the scrapie virus from a filtrate of spinal cord emulsions from sick animals, but such an undertaking required quantities of both money and patience. Assessing the infectious potency of various samples resulting from tests of purification would require the inoculation of a large number of sheep, and the results would not be known until two years later. The two men might have had the patience, but their work came to an end with Cuillé's retirement and Chelle's death. Research on the causative agent of scrapie would barely progress for the next twenty years.

Yet the work of Cuillé and Chelle still marked great progress in the hunt for The Disease. It was as though, sensing danger, The Disease then counterattacked. The offensive was launched in Scotland, around the same time that the two Frenchmen were inoculating their sheep.

The veterinarian W.S. Gordon was studying another ovine disease, louping ill, which was sometimes confused with scrapie.[3] It had been known since the early 1930s that louping ill was caused by a filterable virus that developed in animals' central nervous systems and was transmitted by ticks. Gordon was eager to halt the spread of this disease, which took a substantial toll among Scottish sheep, and set about preparing a vaccine. Here, he took the same approach used by Pasteur for rabies. He gave sheep intracerebral injections of the virus, let the virus multiply for five days, then sacrificed the animals and removed their brains and spinal cords, which contained large quantities of the virus. A suspension was made of nervous system tissues, which were then treated with a solution of formalin (formaldehyde), known to inactivate all viruses. The inactivated viruses lose their infectious nature, but not their ability to trigger a defense mechanism in the animal. The animal's body retains a "memory" of that defensive reaction, which protects it against subsequent infection by the active form of the same virus: The animal has been vaccinated. Suspensions containing the louping ill virus inactivated by formaldehyde proved to be an excellent vaccine against the disease; these were the subject of numerous field tests between 1931 and 1934. Having been determined to be safe and effective, it was manufactured for general use. Three lots were prepared from groups of 140, 114, and 44 sheep, yielding 22,270, 18,000, and 4,360 doses of vaccine respectively. Vaccination took place in 1935, and no complications were initially reported.

Unfortunately, two and a half years after vaccination, in September 1937, two farmers complained that scrapie had appeared in their flocks of blackface sheep—a breed in which the disease had never before been observed. Moreover, the disease affected only animals that had been vaccinated against louping ill in 1935. Painstaking epidemiological study confirmed that the scrapie had been transmitted by the vaccine, and that only lot number two was to blame. When the origin of the 114 animals used to prepare that lot was traced, it proved that although the majority were of the blackface or greyface breeds (hitherto free of scrapie), eight lambs were Cheviots, which were known to be prone to the disease. Although perfectly healthy, those lambs had been in contact with ewes that had subsequently come down with scrapie. There was thus good reason to suppose that one or more of those eight lambs were carriers of the causative agent of scrapie, even though they showed no symptoms of the disease. They were so-called healthy carriers. The infectious agent present in the nervous system tissues of one or more of these lambs had clearly contaminated all of lot number two. The overall contamination rate could not be determined, because the majority of vaccinated animals had been adult ewes and had been sent to slaughter before the disease had had time to appear. On the few farms where the contamination rate could be established, it varied from 1 to 35 percent, averaging 5 percent.

Gordon then became aware of the work of Cuillé and Chelle by reading the one article they published in English, in 1939. He remarked that "it was a curious coincidence that while they were doing their transmission experiments their work was being confirmed by the unforeseeable infectivity of a formalinized tissue vaccine."[4]

Gordon and other British veterinarians learned a lesson from this accident, which could have been catastrophic for sheep farming in the United Kingdom had the contamination not been swiftly detected.

They determined therefore to tackle anew the problem of scrapie. First of all, they had to identify the mysterious infectious agent so that they could prepare a vaccine against it if possible. They did not shy away from the expense of this project: Their experiment used no fewer than 788 sheep. Unfortunately, the outcome would not rise to the level of this expense. The two main results—apart from confirming the work of Cuillé and Chelle—were that it was possible to achieve shorter incubation periods, sometimes as brief as seven months, by employing intracerebral inoculation; and, crucially, that the infectious agent was resistant to formaldehyde, as the vaccine accident had shown. This was a very important point, for no known virus had that attribute.

For a few years, scrapie research seemed to be marking time. The outbreak of the Second World War, of course, had something to do with this. A veterinary problem of minor interest was not high on the list of concerns.

7

AND GOATS, *AND* MICE

THINGS CHANGED IN THE EARLY 1950s. By then scrapie had turned up in Canada, the United States, and Australia in the wake of importation of sheep from the United Kingdom. Those countries, along with New Zealand, imposed an embargo on such sheep unless they could be guaranteed scrapie-free. Now, finding the source of the disease became an economic issue. Motivated and well financed, British veterinarians resumed large-scale experimentation at two major research centers: the Agricultural Research Council's institute in Compton, Berkshire, and the Moredun Research Institute in Edinburgh. Although they were unable to identify the mysterious scrapie virus, they made many important discoveries, three of which I shall discuss here.

The first related to the distribution of the infectious agent in the various organs of a sick animal. It was known to be found in the brain and the spinal cord, but what about other organs? To find out, it was necessary to prepare macerates (crushed tissue preparations) of the various organs of sick animals and see if these could transmit scrapie to healthy animals. Two of the researchers, Iain Pattison and Geoffrey C. Millson, used goats in their experiments. These, in fact, turned out to be far more susceptible than sheep to inoculation; attempts at goat-to-goat transmission were generally 100 percent successful, while the figure was around 25 percent in sheep. Their experiments showed that large quantities of the infectious agent were present in the brain and in the nearby pituitary gland, and somewhat smaller quantities in the cerebrospinal fluid, the sciatic nerve, and the adrenal and salivary glands. Very small quantities were found in muscle tissue, but it could not be detected in the blood or the urine. The infectious agent, therefore, was not confined to the nervous system—the only tissues in which lesions had been observed.

The second discovery also resulted from Pattison and Millson's experiments with goats. Sheep-to-goat transmission by intracerebral inoculation—injection into the brain—resulted in two clinically distinct types of scrapie: "drowsy," which was mainly manifested in neurological symptoms from the outset, and "scratching," whose initial symptoms involved itching before progressing to the neurological variety. If brain tissue from a goat with the "drowsy" clinical type was used to inoculate another goat, this would result several months later in a case of "drowsy" scrapie. Similarly, brain tissue from a goat with the "scratching" type would cause "scratching" scrapie. There appeared to be two strains of the scrapie agent, causing somewhat different diseases. Pattison and Millson offered the theory that these two viral strains were also found among sheep, which could explain the diversity of clinical symptoms that had been observed, as well as the fact that the names for the disease varied by locality—the French had called the disease *tremblante* because their sheep were affected mainly by the "drowsy" strain, while the British had called it scrapie because their sheep harbored the "scratching" virus. Here was yet another example of the many ways in which The Disease could disguise itself. And the question of different strains would arise repeatedly in the course of the hunt. It remains at issue today.

The third discovery I want to highlight was made by a close colleague of Pattison and Millson, Richard Chandler, whose discovery was to expedite research considerably. In 1961, Chandler succeeded in transmitting scrapie to mice. This completely new outcome was published in the British medical journal *The Lancet,* whereas almost all previous work in the field had been published in veterinary journals. As we shall soon see, this was when the hunt for The Disease began to interest non-veterinary physicians.

Chandler was not deterred by the failure of his predecessors in transmitting scrapie to mice; he was determined to try. He was encouraged by two factors. The first was the work of Pattison, Millson, and a few other researchers, who had perfected a method for effectively and reproducibly transmitting the infectious agent in goats. Successive transmissions in goats had, so to speak, stabilized the agent, which now gave rise to relatively consistent symptoms in a time frame that was itself comparatively consistent and quite short: from three to seven months. The second factor had its roots in his own work on the susceptibility of mice to bacterial infections. Now, bacteriologists do not work with mice simply trapped in the wild. For many, many years they have bred a number of different lines of mice originating with wild "founders." The stability of each line is maintained by inbreeding. Thus, the mice of a given inbred line are genetically very similar, while mice of different lines display genetic differences that reflect those between the founders of the respective lines.

Chandler worked with three different inbred lines, and observed a varying susceptibility to bacterial infection. He theorized that they might also have differing degrees of susceptibility to the causative agent of scrapie. He inoculated the brains of mice from each inbred line with extracts of brain tissue from goats suffering from scrapie. He carried out two parallel experiments using brain tissue from goats with the "drowsy" and the "scratching" strains of scrapie, respectively. After an incubation period of seven and a half to nine months, several mice manifested typically neurological symptoms. He wrote:

> The *symptoms* suggested disorder of function of the motor nerves, especially those associated with the hindquarters and tail. The mice stood with their hindquarters lowered close to the ground, and they were reluctant to move. The hind legs were occasionally dragged,

the mice eventually walking with a stiff and rolling gait. The tail
was held in an unnatural manner—stiffly and often to one side.
If curled over a finger, the tail often retained a circular form for
several minutes. When held up by their tails affected mice usually
brought their hind feet together, whereas normal mice usually splay
their hind legs. Some of the affected mice had ruffled coats and
arched backs.[1]

When the mice were sacrificed, examination of their nervous systems
showed lesions that were characteristic of scrapie.

In addition to the strong indication that it was possible to transmit
scrapie to mice, this experiment yielded another important result:
Transmission was successful only with material from the brains of goats
that had the "drowsy" strain, and only in one of the three inbred lines of
mice. The second point recalled the varying susceptibility to scrapie of
different breeds of sheep, as often reported by farmers.

In a 1963 article, Chandler confirmed that he had transmitted scrapie
to mice. He indicated that scrapie could in fact be transmitted from
goats to any of his three inbred lines of mice, but that this was far more
difficult in two lines, where only a tiny fraction of the animals developed the disease, and only after far longer incubation periods (thirteen
to fifteen months, as compared with seven to nine months). Chandler
then set about effecting mouse-to-mouse transmission, which he did
without difficulty. But here, the infectious agent acquired two new
properties. First, it had somehow adapted to its new host, bringing on
the disease more quickly, in only four to five months. And second, it developed with equal speed and effectiveness in all three inbred lines of
mice. The strain of scrapie that had adapted to mice was in some way
different from the original strain that had adapted to goats. The
changes were seen too when Chandler successfully retransmitted the
disease from mice to goats after several mouse-to-mouse transmissions.
That success was proof positive that the disease observed in mice was indeed scrapie. As noted, however, only the "drowsy" type could be trans-

mitted to mice. But with transmission in the opposite direction, goats inoculated with the infectious agent from mice with "drowsy" scrapie
sometimes came down with the "drowsy" type and other times with the
"scratching" type—and sometimes with a mixture of the two. Such a
change had never been observed in goat-to-goat transmission.

It was thus learned that scrapie could be transmitted between species
as different as goats and mice; that transmission could be more or less
difficult depending on the genetic traits of the animals concerned; and
that once transmission had occurred, the agent adapted to its new host
by acquiring new properties.

By using mice, Chandler was able to do a series of basic experiments
whose cost and duration would have been prohibitive using goats or
sheep. For example, he carried out measurements of the infectious
agent; previously, scientists had simply used undiluted extracts of various tissues for inoculation, and the result was basically all or nothing—
either the inoculated animal developed scrapie or it did not. Of course,
in those previous experiments the incubation period and the percentage
of animals that contracted the disease provided some idea of the quantity of infectious agent in the inoculation, but this was very imprecise.
Chandler, who was able to use a virtually unlimited number of animals,
prepared a series of different concentrations of tissue extracts and used
them to inoculate a great number of mice. Using an extract of mouse
brain, he observed that the more dilute the extract, the longer the incubation period. But even with a 1:100,000 extract (equivalent to one
tablespoon of extract in more than 390 gallons of water), he succeeded in
causing the disease in less than six months in all inoculated mice. And
some mice contracted scrapie eight to nine months after inoculation
with a 1:100,000,000 extract (one tablespoon per 390,000 gallons—the
equivalent of a typical Olympic-sized swimming pool). Clearly, there
is a considerable quantity of infectious agent in the brains of sick
mice. Considering that a mouse can be inoculated with about one one-
hundredth of a cubic centimeter (0.00034 fluid ounce), it is clear that
extracts from the brain of a single infected mouse would suffice to con-

taminate millions or even billions of other mice.

Chandler also studied the effect of different methods of inoculation: injection into the brain (intracerebrally) and into the spinal cord, through the peritoneum (intraperitoneally), subcutaneously, and orally, using a gastric tube. In all cases it proved possible to transmit the disease, although with only partial success using oral inoculation; in those cases only about half the animals had sickened after nine months, while all others had shown symptoms in less than seven months.

Mice would now supplant sheep and goats in scrapie research. The fact is that any scientist seeking to purify the causative agent (that is, to separate it from the numerous particles and molecules present in any cell extract or bodily fluid) needed to inoculate hundreds or thousands of animals in order to measure the infectious potency of various samples. This was possible using mice, but less so with sheep or goats. So we will soon be able to shift our attention away from those friendly farm animals. But first, we have to return to the thorny question of how scrapie is transmitted in nature.

8

SCRAPIE IS CONTAGIOUS

THE WORK OF CUILLÉ AND CHELLE, now fully confirmed by British researchers, indisputably established that scrapie was transmissible. Once introduced into an animal's body, the causative agent multiplied, so that after the proper incubation period the tissues of the animal could be used to infect other animals, and so on. But was scrapie contagious? Was it transmitted spontaneously from animal to animal?

MacFadyean's observations seemed to demonstrate that it could be: Think of Mr. A (introduced in Chapter 4), whose flock had been contaminated by ewes bred by Mr. X. Many similar observations had been published in the scientific literature. Yet there remained room for doubt. It was always hard to be sure whether, had they been isolated, the animals thought to be contaminated by other animals might not have contracted scrapie anyway. Such doubts were fueled by the nearly constant failure of attempts to observe contagion in controlled conditions. This problem too was tackled by British veterinary scientists.

First of all, here are the results of experiments published by Iain Pattison in 1964: Seventeen Cheviot sheep (a breed susceptible to scrapie) were kept in a shed for fifty-five months in close contact with a series of sheep and goats infected with scrapie that had been transmitted to them by inoculation. None of the Cheviots contracted scrapie. Nor did the disease appear in any of the 192 goats similarly kept in contact with animals with laboratory-transmitted scrapie. Finally, thirty-three kids—the offspring of three male and twenty-seven female goats inoculated with scrapie at the time of conception—were kept in close contact with their parents, including suckling by their mothers, and showed not the least sign of scrapie four years after birth. How could these negative results be reconciled with the many observations of contagion in normal farm conditions?

Pattison, who worked at the Agricultural Research Council's institute in Compton, undertook a new series of experiments in collaboration with colleagues from the Moredun Research Institute in Edinburgh. This time, the outcome was more conclusive. In two experiments, seventeen goats, from birth, were kept in close contact for extended periods with sheep that had contracted scrapie by natural means. Ten of them came down with the disease. Moreover, three cases were observed in blackface sheep that for about four years had been kept in close and sustained contact from birth with sheep of various breeds that had natural scrapie. Blackface sheep were supposed to be completely invulnerable to scrapie in natural conditions. This was thus an unambiguous case of contagion.

Why did these two series of experiments have such contradictory outcomes? Could there have been uncontrolled differences between the facilities at Compton, where the first series had taken place, and those at the Moredun Institute, which was the site of the second? There could have been other reasons as well. Specifically, the animals used in the attempt to contaminate the others had contracted scrapie experimentally (by inoculation) in the first instance and naturally in the second. Perhaps the two types of scrapie differed in their ability to be transmitted by contagion.

In any event, contagion was observed in the second experiment, which confirmed the many field reports from farmers and veterinarians. Yet, not everybody was convinced. One of the top scrapie specialists, H.B. (James) Parry, continued to believe until his death in 1980 that

scrapie could be transmitted only by inheritance. Furthermore, in an excellent book published as recently as 1998, two other specialists, Ros-
alind Ridley and Harry Baker, expressed doubt that scrapie was
transmitted within flocks, and defended the notion of its exclusively genetic origin.[1] Still, the great majority of researchers endorsed the idea
that scrapie was a contagious disease, even if the contagion is weak. Exactly how the contagion occurred, however, remained to be discovered.

How did the "virus" enter animals in natural conditions? One common entry point for infectious agents is the mouth. Richard Chandler
had shown that oral contamination, although inefficient, was possible in
mice. What about sheep and goats? Pattison and Millson had asked that
very question in 1961, even before Chandler began his experiments. By
feeding sheep and goats a drink containing crushed matter from the
brains of sick animals, they succeeded in transmitting the disease. Out of
fifty sheep of various breeds that had been fed this strange brew, seven
contracted scrapie in the following eleven months. So oral contamination was possible. But was it responsible for natural contagion?

It was not until two decades later, in 1982, that a group of American
researchers found evidence to support that theory. William Hadlow
and his colleagues studied a group of Suffolk sheep belonging to a flock
greatly affected by scrapie, their aim being to learn the distribution of
the infectious agent in different tissues and to see how this varied according to the animals' ages. They sacrificed animals in various age
groups with a view to finding the agent during the disease's development phase and seeing where that agent resided.

In fourteen lambs under eight months of age, they found no trace of
the infectious agent in the tissues they analyzed when they inoculated
mice with those tissues. On the other hand, they detected it in eight out
of fifteen lambs aged ten to fourteen months. In these cases it was present only in the intestine and in the lymph nodes; most of the infected
lymph nodes were those near the pharynx and the intestines.

In a group of three animals aged twenty-five months, the infectious
agent was found in one ewe. It was located in the digestive tract, including

the colon, and throughout the lymphatic tissues, and was beginning to make its appearance in the nervous system, although in low concentrations. No nervous system lesions were visible at this stage, and the animal displayed no clinical symptoms.

Obviously, the infectious agent was found in the nine sheep, aged thirty-four to fifty-seven months, that displayed symptoms of scrapie. The highest concentration was found in the nervous system, although it also infected other organs, including the intestines. And finally, no sign of the infectious agent was found in seventeen adult sheep that displayed no symptoms—although it might have been possible that the disease was in its incubation period.

So, given that the digestive tract and nearby lymph nodes were the first to be affected, infection seemed to take place orally. It then spread to the nervous system via the lymphatic tissues. Infection was more likely among young animals; the infectious agent was found in half the lambs aged ten to fourteen months but in no healthy animal older than thirty months. Moreover, its presence in the intestines throughout the course of the disease suggested that the agent would be found also in the stools, which would then play a role in transmission. That hypothesis has yet to be proved, however; all attempts to find the infectious agent in fecal material have thus far failed.

But Pattison and his colleagues formed another theory. A year or two after their experiments demonstrating the contagious nature of the disease, they realized that the placentas of sick or infected ewes could be a vehicle of contamination. Placentas expelled during parturition are eaten by other sheep, and sheep are generally kept together in a confined space during lambing. Such conditions would be perfect for spreading the disease if contamination could take place orally, which had already been demonstrated, and if the placenta contained the infectious agent, which remained to be proven. No sooner said than done— although the concept of "no sooner" has to be seen in the context of the time it took to carry out the necessary experiments. Anyway, three years later, in 1972, the theory had been substantiated: Oral inoculation of a suspension of placental membrane from a sick ewe resulted in the appearance of scrapie both in sheep and in goats. If the placenta played a major role in natural transmission, this could be another explanation for the apparent contradiction between the two series of experiments by which Pattison had tried to prove that the disease was contagious: In the first series, which had a negative result, the scrapie carriers were either males or nonpregnant females; but there were some pregnant females in the second series, in which contagion was observed.

A possible route for contamination had thus been established, but what was its real importance? And were there other possible routes? No one knew.

Pattison felt that the placental theory could even account for a strange observation made in 1954 by the Icelandic virologist Björn Sigurdsson and confirmed by others: Healthy animals could be contaminated without direct contact with sick animals, merely by frequenting places where such animals had spent time. That recalled the old "accursed" fields in which animals regularly contracted anthrax, for which Pasteur had found an explanation (see Chapter 3). In Iceland, animal husbandry had always been a major economic activity. Ovine diseases, often introduced when the animals were imported from abroad, had more than once caused famine on the island. In the early 1940s such a situation was brewing, as three new diseases were devastating the sheep population. It appeared that they had been introduced in 1934 by twenty sheep purchased in Germany.

Having returned from advanced training in Denmark and the United States, Sigurdsson tackled the new diseases. He found that one of them was caused by a previously unknown virus, later named the visna virus. To eradicate that virus, which was causing severe damage to Icelandic farmers, it was necessary to systematically destroy all flocks containing affected sheep. Some of those flocks also included animals infected with *rida,* the Icelandic equivalent of scrapie. A few months or a few years after such a flock was destroyed, it would be replaced with animals from areas free of *rida.* Without fail, the new flock would contract *rida.* But if animals of the same origin were used to replace a flock that had been free of *rida,* the disease would not appear in the new flock. It appeared as though the fields or sheepfolds had been contaminated with the *rida* agent, and as though this agent remained until the arrival of the new animals and then infected them.

It was Pattison's view that, in light of its great hardiness, the scrapie agent could persist in pastures or farm buildings once having been deposited there by means of contaminated placentas. Indeed, experiments by other scientists, published in 1991, confirmed that extracts from the brains of animals sick with scrapie retained a significant degree of infectiousness after having been buried in a garden for three years. Perhaps, then, animals could become infected by eating vegetation that had previously been contaminated via placentas.

A great number of animals had to be studied in experiments to confirm the infectious and contagious nature of scrapie. In the course of that work, lesions caused by the virus were studied far more closely than Charles Besnoit had been able to do in the late nineteenth century. This work involved sheep, goats, and mice alike, and it confirmed that no non–nervous system organs underwent visible change. On the other hand, nervous system lesions proved to be more extensive than Besnoit's observations had suggested. Besnoit had observed changes principally in the spinal cord and the peripheral nerves, but in fact the brain itself was the focus of large-scale degeneration of the neurons, which are the basic cells of the nervous system. In the brain, as in the spinal cord, the degeneration was characterized in particular by the presence of bubble-like vacuoles both within and between cells. Some parts of the brain resembled Swiss cheese, or a sponge.

By the early 1960s, knowledge of scrapie had expanded greatly. We knew that it was an infectious, moderately contagious disease. And, while the causative agent had yet to be identified, we knew that it had very distinctive characteristics (such as its resistance to formaldehyde) and that it could multiply in a variety of tissues, although it had a preference for nervous system tissues, where it caused characteristic lesions.

These results—obtained by a handful of mainly British and French veterinary scientists at the cost of the lives of many, many animals— were known only to other veterinary scientists and well-informed farmers, who were the audience for publications and conferences on the subject. Few physicians had even heard of the disease, and even fewer members of the lay public.

But in the course of the 1960s, the Fore people of Papua New Guinea would break down the walls surrounding knowledge of scrapie, for The Disease had struck them, this time under the name of kuru.

9

KURU AND THE FORE PEOPLE
OF PAPUA NEW GUINEA

SOME PLANTS AND ANIMALS ARE so-called living fossils. They seem to have survived since time immemorial while their contemporaries fell by the wayside, yielding to new species better adapted to a changing world. As of the mid-twentieth century, the human species too included its share of well-adapted survivors: groups cut off from the rest of the world, who lived as our distant ancestors had lived. Among them were a number of ethnic groups on a huge island to the north of Australia, which had been "discovered" in the sixteenth century by Portuguese and Spanish navigators, who named it New Guinea. Unknowingly, and to its great detriment, one of these groups—the Fore—was to add a new dimension to the hunt for The Disease.

For a very long time New Guinea was terra incognita. It became clear that it was an island only in the late eighteenth century, thanks to the voyages of Captain James Cook, and exploration of the interior would not begin until early in the following century. This beautiful island, however, had a dark reputation. Even the indomitable Captain Bligh, desperately short of food and water following the *Bounty* mutiny, did not dare venture there for fear of local tribes reputed to be bloodthirsty cannibals. Geography and climate also posed formidable obstacles to exploration. The mountainous island, with peaks as high as thirteen thousand feet, was covered in dense tropical forest, and visitors from temperate zones found it difficult to stand its extremely hot and humid climate.

The island is about four hundred miles across at its widest, and some fifteen hundred miles long; today it is divided into two more or less equal parts. The western half is the Indonesian province of Irian Jaya, and the eastern half is the young nation of Papua New Guinea, which gained its independence in 1975. In the 1950s, however, Papua New Guinea was still under Australian administration. In those days, some barely accessible mountainous areas were home to populations still living in the Stone Age. They were armed with bows and arrows, they used stone axes, and they plowed the land with sticks. They knew nothing of metals or even the wheel. As among all primitive peoples, magic and the supernatural played a key role in their lives. Sorcerers were the depositaries of supreme power, and intimacy with the dead was the rule. The regions in which these populations lived had hardly ever been visited by people of European origin.

The Australian administration was determined to gain control of these primitive peoples—to impose its authority, to put a stop to wars between villages, and to eradicate cannibalism. Among those sent by the Australian government was Vincent Zigas, an enthusiastic young German doctor of Estonian origin, who found himself in one of the areas being pacified.

Zigas arrived in New Guinea in 1950. In a fascinating book describing his extraordinary adventure, he questioned what initially made him leave his home country: "Was it because I was looking to evade the fearful new battle I saw looming in my Vaterland, the battle between the star-spangled banner and the hammer-sickled emblem? Or was I seeking a place of tranquility where I could study humanity, people close to nature, and offer my services to help fight disease and its incapacities? Or was it perhaps because of my desire to study the people—the people of yesteryear[?]"[1]

After four months of training in Sydney, during which he learned everything he needed to know about the territories in which he was to work, Zigas's first assignment was in New Guinea. The purpose of his assignment was to help people fight the many diseases that thrived in the region. (His success earned him Australian citizenship and a job with the Australian Public Health Service.) In 1955 he was transferred to a mountainous region in central New Guinea. There were few volunteers for self-exile in such isolated spots, and Zigas was the only physician in a very populous area. He had a tough job to do—not unlike that of today's humanitarian doctors. After months of hard work, he suffered an accident and had to undergo knee surgery in a little nearby town. At a social gathering during his convalescence, he met a military officer whose mission was to pacify a sector inhabited by an ethnic group known as the Fore. Zigas asked about the health situation among that population, and the officer mentioned a number of pathologies that were common in the region. Then he mentioned kuru.

The officer had become aware of kuru very early in his pacification mission. His diary entry for December 6, 1953, reads in part: "Proceeding SW across range, and down and across a small creek ascending to Amusi villages, nearing one of the dwellings, I observed a small girl sitting down beside a fire. She was shivering violently, and her head was jerking spasmodically from side to side. I was told that she was a victim of sorcery, and would continue this shivering unable to eat until death claimed her within a few weeks."[2]

He later encountered a number of similar cases. The ailment was called kuru, a Fore word meaning "to tremble with fear or cold." Zigas seemed interested; the officer invited him to pay a visit and promised to send a guide.

Things move slowly in the hostile, mountainous jungle where the indigenous peoples lived. Communications were not easy, and the inhabitants had eternity before them. So it was three months before the officer's guide arrived to pick up Zigas, in September 1955. They set out immediately. After two days of walking in the mountains, they arrived in a little hamlet, nothing more than a handful of huts. The guide pointed to one of these, and Zigas entered. A woman about thirty years old was sitting in the corner. She had a very strange look about her. She did not actually look ill, but she was emaciated; her eyes were blank and her face seemed mask-like. Every so often her body would tremble slightly, as though from the cold, even though the temperature was quite hot. Zigas was told that the woman was very sick, that she had been bewitched. He tried to lift the enchantment, using what medical tools he had at his disposal, but was completely unsuccessful. The other villagers were not surprised. What could this pale stranger do against the omnipotence of a sorcerer? A little later, the guide said to him, "Dokta, don't use your magic medicine any more. It will not win [against] our strong sorcery. In my place you will see plenty people dying from this."[3] Zigas had just encountered his first case of kuru.

Two days later, he reached the center of Fore territory and met his officer acquaintance, who seemed to think that kuru was a form of mass hysteria. The Fore believed that no death was natural. When a family member or friend died of disease, they would search for the sorcerer responsible for it. If there was a known sorcerer in the area, he would be the prime suspect. If not, suspicion could fall on anyone—a personal enemy or someone with an unusual appearance or lifestyle. Such enchantment demanded vengeance: a form of ritual murder called *tukabu*. As a rule, then, every death from kuru involved a second death from *tukabu*.

Zigas provides careful and poignant descriptions of many of the cases he encountered. For instance, he wrote of a woman seated outside her doorway: "She was holding on her lap a limp figure, grossly emaciated to little more than skin and protruding bone, the shivering skeleton of a boy, looking up at me with blank crossed eyes."[4] It was the woman's only child, and he died the next day. Her husband had suffered death by *tukabu*.

On another occasion, Zigas's attention was drawn by a young boy carrying on his shoulders a long bamboo tube full of water. The lad seemed to stumble as he tried to go through the narrow opening in a fence. Zigas was surprised because Melanesian children are generally very agile. His guide explained that the boy's legs were weak because he had kuru. The boy managed to get through the opening, and the other children burst out laughing. The water carrier himself joined in the merriment. But to Zigas, his laugh sounded debilitated and foolish.

Zigas was soon able to formulate a clinical picture of the disease, which was remarkably consistent in its development. The first symptoms were problems in walking and balance, then in the carriage of the arms and torso, gradually leading to total disability. The limbs, body, and neck displayed tremors similar to those resulting from a psychic shock; these diminished when the victim was at rest, and ceased during sleep. Associated symptoms included pronounced cross-eyedness and great emotional instability. The patient became incapable of moving about and ended up prostrate and bedridden before falling into a coma, then dying. Intellectual powers seemed to be barely affected, although patients quickly lost the ability to speak.

This clinical picture puzzled Zigas. He came up with several theories, including that of the officer: autosuggestion or mass hysteria. Because they were convinced that they had been bewitched, victims really fell ill. For nearly a year, he spun theories and tried to interest Australian researchers and administrators in the problem, with mixed success. Although he secured an appointment with the famous scientist Sir Frank Macfarlane Burnet, director of Melbourne's Walter and Eliza Hall Institute of Medical Research, Burnet showed only polite interest. In contrast, a great virologist at the same institution, Gray Anderson, was far more interested and was prepared to help. After a second trip to Fore territory, from October 22 to November 12, 1956, Zigas brought him twenty-six blood samples and a brain from kuru victims. It was hoped that Anderson could find some infectious agent, perhaps a virus, in these samples. A few weeks later, the results came back, and they were negative: None of the usual virological methods could detect any infectious agent whatsoever. The mystery grew more obscure.

On the eve of his return to Fore territory, scheduled for March 14, 1957, Zigas received a strange visitor, whom he described in these words:

At first glance he looked like a hippie, though shorn of beard and long hair, who had rebelled and run off to the Stone Age world. He wore much-worn shorts, an unbuttoned brownish-plaid shirt revealing a dirty T-shirt, and tattered sneakers. He was tall and lean, and one of those people whose age was difficult to guess, looking boyish with a soot-black crewcut unevenly trimmed, as if done by himself. He was just plain shabby. He was a well-built man with a remarkably shaped head, curiously piercing eyes, and ears that stood out from his head. It gave him the surprised, alert air of someone taking in all aspects of new subjects with thirst.[5]

The newcomer was D. Carleton Gajdusek, who in 1976 was to share the Nobel prize in physiology or medicine.

In 1957, Gajdusek was thirty-seven years old. He was an American pediatrician with solid scientific training gained in the labs of a number of prominent researchers. He was in Australia as a guest researcher at Frank Macfarlane Burnet's laboratory at the Hall Institute. His lab work achieved important results, but he also took advantage of his time in Oceania to study child development and childhood diseases in primitive societies. That was why he had decided to stop in New Guinea on his way back to the United States. At the time of his arrival in Port Moresby, the administrative capital of the territory of Papua New Guinea, he had never even heard of kuru. Sir Mac, as Burnet was familiarly known, had never breathed a word of it to him—which angered Zigas when he learned of it. So it was the newly appointed director of public health in the territory who filled him in on Zigas's observations. Gajdusek was intrigued and immediately decided to join Zigas. Hence his unexpected appearance on the eve of Zigas's departure on a further expedition to Fore territory.

Delighted that he had at last found a research scientist who wanted to take a close look at the problem on which he had been focusing for a year and a half, Zigas brought Gajdusek along with him. That marked the beginning of a year of intensive work and of an effective and friendly collaboration not just between those two men but among a number of others as well, both Australians and indigenous Papua New Guineans who supported the research. A makeshift laboratory was set up, along with accommodations for patients. Blood and other samples could be taken and autopsies and some basic analyses could be carried out in acceptable if not ideal conditions. Some samples were sent to Australia or the United States for analysis. Work in the lab alternated with expeditions intended to flesh out the clinical picture, to assemble epidemiological data, and to pinpoint the geographical distribution of the disease.

September or October 1957 brought the results of Igor Klatzo's histopathological analyses (microscopic examination of tissue sections) carried out in the United States at the National Institutes of Health. Although these did not answer the questions that Zigas and Gajdusek were asking about the origin of the disease, they included facts that were subsequently to be of great importance.

First of all, Klatzo observed generalized neuron degeneration in the brain and spinal cord. In his view, the only other disease that caused the same changes was Creutzfeldt-Jakob disease. Neither Gajdusek nor Zigas was familiar with CJD, of which only a score of cases had been described, none of them in English-language journals (although German, Zigas was a humanitarian physician who would not have kept up with the scholarly literature).

And in half of the twelve cases he studied, Klatzo noted the presence of unusual marks—"plaques"—mainly in sections prepared from the cerebellum but also in other nervous system tissues. Using the appropriate stains he visualized the structure of these plaques, which generally seemed to consist of a structure of fibers organized around a dark center. The plaques bore a certain resemblance to the so-called senile plaques found in the brains of patients suffering from Alzheimer's disease, although they differed in a number of details. Similar plaques had not been described in CJD patients, apart perhaps from a single case. And in that case the distribution was different, for the plaques seemed to be absent from the cerebellum.

About the same time as those results arrived, Gajdusek began to write two articles that would inform the scientific community of the observations he and Zigas had made. These appeared in November 1957 and described in detail the clinical picture and set out the initial epidemiological data:

The illness runs an afebrile course and is characterized by the insidious onset of ataxia, which becomes progressively more severe and is soon accompanied by a fine tremor involving the trunk, head and extremities. Both involuntary tremor and ataxia increase, . . . with exaggeration during voluntary motor activity or fatigue, subsidence during rest and disappearance during sleep. . . . No evidence of systemic disease, particularly liver involvement, can be found.

Involuntary tremors, ataxia and incoordination continue to increase in severity for one to three months from their onset . . . by which time the patient usually requires the support of a stick for walking. A month or two later the patient is no longer able to walk or stand at all without considerable support, and equilibrium in the sitting posture is soon thereafter impaired. Intelligence remains normal during the early months of illness, but speech slowly becomes blurred and slurred and finally no longer intelligible. Together with this dysarthria, a progressive slowing of intellectual functions is apparent. The patients often display a marked emotionalism, with excessive hilarity, uproarious, foolish laughter on slight provocation, and slow relaxation of emotional facial expressions. . . . [In] general, the patient remains well integrated in his social setting until he slowly falls into greater and greater incapacitation. When he is no longer ambulatory in the native setting he is usually left to die in the low, dark kunai-grass habitations and seldom if ever is carried out into the sunlight. Urinary and fecal incontinence develop, and speech is completely lost. . . . Finally, swallowing and chewing are no longer possible, and the helpless patient succumbs with rapid starvation, decubitus ulcerations and terminal static bronchopneumonia. . . . A convergent strabismus is a nearly universal late development in the disease.

Kuru rarely lasts much over a year, and is often fatal within three to six months.[6]

Kuru never appeared in people younger than age four. Among children, boys and girls were equally affected, while, in adults, ten to twenty times more women than men contracted the disease. That was odd, and it needed to be explained. Kuru's geographical distribution was surprising as well. It was seen only in a small area of about twenty by forty miles. Only the Fore people, and to a lesser extent neighboring groups with which Fore had married, were affected. In the area in question, an average of 1 percent of the population was affected, and about 1 percent died of kuru every year. In some Fore clans, 5 to 10 percent of the population was affected, and half the deaths over the preceding five years had been attributed to kuru. The total number of kuru deaths was several thousand. To those already high figures had to be added the deaths of sorcerers in vengeance killings—*tukabu*—and of young children of women who died of the disease.

As to the cause of kuru, the authors were forced to admit that they did not know. When Gajdusek first heard of the disease, an infectious origin is what came to mind. And his first studies were aimed at finding the causative agent. But as work progressed, that theory seemed increasingly untenable. The most surprising thing was the lack of an immune system reaction in patients. In principle, every infection triggers the body's defense mechanisms. This is reflected by such phenomena as increased body temperature and inflammatory signs indicating the mobilization of specialized cells designed to eliminate the intruder. But none of this was seen in kuru patients. Then there was the repeated failure to detect an infectious agent in samples sent to Australian and American labs. On the other hand, the fact that the disease was found within families suggested a genetic origin, perhaps combined with an environmental factor that was specific to this ethnic group.

10

THE WALL COMES DOWN

NOT THE BERLIN WALL—this was 1959, and that structure would not rise for another two years—but the wall that segregated physicians and veterinarians was toppling. That year saw the first suggestion of a similarity between a human form of The Disease—kuru—and its animal form, scrapie.

Let us return to the Compton research center in England, where Iain Pattison and his colleagues were at work. When the issue of scrapie in sheep that were exported to the United States arose, an American veterinary scientist, William Hadlow, was sent to Compton in 1958. On September 5, 1959, Hadlow published a brief letter in *The Lancet* that was to be of crucial importance. It began very modestly—as befitted a veterinary scientist publishing in a journal of human medicine—with the author acknowledging that it was risky to draw too-close comparisons between human and animal diseases. It went on, however, to offer a convincing list of similarities between scrapie and kuru. And it ended with a humble proposal: "[It] might be profitable, in view of veterinary experience with scrapie, to examine the possibility of the experimental induction of kuru in a laboratory primate, for one might surmise that the pathogenic mechanisms involved in scrapie—however unusual they may be—are unlikely to be unique in the province of animal pathology."[1]

The similarities between scrapie and kuru were such that it would be wise to consider whether, like scrapie, kuru was transmissible. Human experiments were out of the question, but transmission could be attempted in our closest relatives, the primates. After all, chimpanzees are no more distant from humans than goats are from sheep. And in his letter Hadlow recalled that the incubation period could be extremely long, which would require patience in anyone who wanted to try to transmit kuru.

Hadlow's suggestion was a revelation for Gajdusek, who had temporarily left Papua New Guinea with the eager encouragement of the Australian authorities, who were not keen on an American commandeering the research on kuru. Gajdusek wanted to learn more about scrapie, and paid a visit to Hadlow and his fellow veterinary researchers at Compton and at the Moredun Research Institute. He came away convinced that it was important to try to transmit kuru to laboratory animals, especially primates. And he understood that he would have to be patient. He inoculated his first chimpanzees in August 1963.

While we wait to see what happened, let us go back to Oceania. A young Adelaide doctor, Michael Alpers, had arrived in Fore territory to study kuru with Gajdusek, who was again spending time in the region. Alpers and Gajdusek began an extensive epidemiological study of the disease focusing in particular on how its impact had evolved over the years. One of their first findings was that kuru seemed to be a recent phenomenon. Numerous testimonies agreed that it had appeared only forty or fifty years earlier, around the turn of the twentieth century. Another very perplexing finding was that the epidemic had receded since 1957, when Zigas and Gajdusek had made their first observations. The decline was limited but significant, especially among children. It was very tempting to correlate this with the pacification process, which had been launched in that area just as the first studies on kuru were being carried out. That theory was bolstered by the observation that the steepest decline in the number of cases was observed in areas where European civilization had penetrated earliest. Because contact with Europeans had brought many changes to indigenous ways of life, it was hard to know which of those changes had affected the incidence of kuru. Among the customs that the pacification patrols had totally eradicated was that of ritual cannibalism, in which the Fore ate deceased members of their families. Furthermore, for the Fore people, eating the corpses of those who had died of kuru was supposed to protect a person from the disease.

Getting back to our—or rather Gajdusek's—chimpanzees, Gajdusek and his colleagues Joe Gibbs and Michael Alpers published their first outcome in February 1966. Within eighteen to twenty-one months after inoculation, three chimpanzees that had been given intracerebral inoculations of suspensions of brain tissue from three different people who had died of kuru displayed symptoms astonishingly similar to those of human kuru. They were all there, to a degree that varied from chimp to chimp: loss of balance, tremors, inability to feed, cross-eyedness, and, within a few months, death. The nervous system lesions too were found to be very similar, with the typical degeneration of the cerebellum. Gajdusek had reproduced, for kuru, what Cuillé and Chelle had done thirty years earlier for scrapie. And Hadlow's prediction turned out to be accurate: Kuru behaved very much like scrapie.

Gajdusek, Gibbs, and Alpers confirmed that outcome by successfully transmitting kuru to several other chimpanzees, not only from human victims but also from the chimps to which they had initially transmitted the disease: a second passage. In the second passage, the incubation period fell to one year. This reflected adaptation by the "virus" to its new host, as had been seen in the transmission of scrapie between species.

Gajdusek then began to wonder whether it would be possible similarly to transmit other chronic nervous system diseases, but his first results—involving diseases such as multiple sclerosis and Parkinson's disease—were all negative. Then a second attempt, in 1968, succeeded in transmitting Creutzfeldt-Jakob disease to a chimpanzee. Symptoms in that animal were similar to those observed in chimps to which kuru had been transmitted. CJD, kuru, and scrapie thus appeared to be related diseases caused by agents that were most likely similar.

At that point a generic term started to be used to describe this group of diseases. The Disease now had a scientific name: subacute spongiform encephalopathies. "Subacute" because these diseases developed slowly; "spongiform" because they caused parts of the brain to resemble a sponge; and "encephalopathies" because they affected the encephalon—commonly known as the brain. These diseases came later to be defined also as transmissible, hence the term TSSEs: transmissible subacute spongiform encephalopathies.

Back in Papua New Guinea, Alpers was pursuing his epidemiological studies and confirmed that the trend observed in 1964 had accelerated. Kuru was disappearing, first and foremost among children, and it became increasingly obvious that this had to be attributed to the cessation of cannibalism. The work of Gajdusek and his colleagues had established that the disease was infectious, which made transmission through cannibalism very probable. Moreover, it was known that women and young children were far more likely than men to be exposed to the infectious agent. Women had the job of butchering the corpses, during which they could be contaminated through small cuts in their skin. Such contamination would not have spared very young children in their care, who would have been present during these activities. Furthermore, only women and children ate the brains and viscera; the "good parts" (muscle tissue) were reserved for the men. These practices could have accounted for the unusual epidemiological character of kuru, a disease that almost exclusively affected children of both sexes and adult women. The fact that the disease seemed to run in families was easily explained by the fact that cannibalism was practiced within the family circle. As to whether contamination took place orally or through cuts in the skin, Gajdusek always inclined toward the latter theory. His reasoning was not always clear—in 1961, for instance, Pattison and his colleagues had succeeded in transmitting scrapie orally in both sheep and goats—but was probably based on the fact that considerable data indicated very early contamination in young children, at an age when they would not yet have participated in ritual mortuary meals.

Kuru has now more or less disappeared, but study of the remaining patients provides data on its incubation period. Because cannibalism had completely ceased by the late 1950s, all subsequent cases of kuru implied contamination before that time. Cases of kuru have continued very occasionally to appear four decades later, at the turn of the twenty-first century. Insofar as kuru and Creutzfeldt-Jakob disease would prove to be very similar—nearly identical, in fact—that information would be very important with respect to cases of accidental transmission of CJD, as we shall see later.

But how did kuru first come to the Fore people? We can only speculate. The most likely conjecture is that a Fore contracted a case of CJD of unknown origin, as has occurred all over the world, and that the infectious agent spread among the population through cannibalistic practices. The successive passages from Fore to Fore could have enabled the infectious agent to adapt, thus producing kuru's characteristic clinical features, which differ somewhat from those of Creutzfeldt-Jakob disease.

Kuru had indeed released scrapie from its veterinary ghetto, but it did much more besides. Study of kuru demonstrated for the first time that human diseases causing nervous system degeneration could be infectious. The causative agents in both humans and animals seemed to be highly unusual, in terms of their slow development in the infected host as well as their physical characteristics. This strangeness soon drew the attention of a new generation of tenacious researchers. In the meantime, however, Gajdusek's success—his extraordinary feat in the Papua New Guinea mountains and his proof that kuru and, subsequently, Creutzfeldt-Jakob disease were caused by an infectious agent—deserved recognition by the scientific community. This would come a few years later, in 1976, when the Nobel committee awarded him its prize in physiology or medicine (jointly with Baruch S. Blumberg).

But alongside this respect for the achievements of Gajdusek and his team, concerns began to emerge. If spongiform encephalopathies were transmissible not only within but between species, if oral contagion was possible, and if the causative agent of scrapie was similar to that of human diseases, could someone get scrapie by eating lamb or mutton? This would have been in Gajdusek's mind when he tried to transmit scrapie to primates and, conversely, to transmit kuru and CJD to various animals other than chimpanzees, including sheep and goats. After numerous failures, he finally succeeded in 1972—five years and five months after inoculation—in transmitting scrapie from a mouse to a rhesus monkey. He had in the meantime transmitted both kuru and CJD to a large number of species of primates, including rhesus monkeys, which proved that a given species could be susceptible both to scrapie and to human diseases of the same type, and that the resulting symptoms were similar. Eight years later, in 1980, Gajdusek's team would report oral transmission of kuru, Creutzfeldt-Jakob disease, and scrapie to small monkeys (squirrel monkeys). Here, transmission was "natural" in the sense that the contaminated tissues were merely mixed into their food.

But we are getting ahead of ourselves, so let us return to the late 1960s. By now, the group of TSSEs—transmissible subacute spongiform encephalopathies—had been clearly defined. They affected both humans and animals and were characterized by very similar clinical symptoms and by easily recognizable nervous system lesions. They could all be easily transmitted by intracerebral inoculation with infectious matter, as well as by other routes, including orally. The big question now was: What were the infectious agents? Scientists were talking of viruses, but these were very strange ones. Although their existence had been postulated by Cuillé and Chelle as early as 1936, no one had yet seen one. Were these really viruses, or a new kind of infectious agent that had still to be described? A science newly emerging in the 1950s and 1960s would tackle this problem: molecular biology.

11

FROM PEARL NECKLACE TO DOUBLE HELIX

IT WAS THE LATE 1960s. In Paris, the United States, and throughout the developed world, students were taking to the streets to protest war, intolerance, and other social ills.

How was the hunt for The Disease going? A composite sketch was emerging: In very similar guises, it attacked sheep and goats as scrapie; it affected the Fore people of Papua New Guinea as kuru; and it threatened the rest of the human world as Creutzfeldt-Jakob disease. It could be transmitted experimentally both within a species and between species, so it had to be caused by an infectious agent. In nature, however, it was only moderately contagious. Contagion had been observed among sheep and goats, but the mechanism was poorly understood; it could be through the consumption of infected placentas, but there may have been other routes as well. In humans, the cannibalistic practices of the Fore had turned kuru into an epidemic, but Creutzfeldt-Jakob disease appeared to be completely noncontagious, since no case of human-to-human transmission had been reported.

But two sets of facts created some uncertainty about that composite sketch.

The first included a great number of observations suggesting that The Disease could be hereditary. The theory that scrapie was of genetic origin in sheep, offered by Thomas Comber as early as 1772, still found advocates, such as the respected veterinary scientist H.B. (James) Parry, who refused to believe that it was contagious. And in humans too, the existence of multiple cases of Creutzfeldt-Jakob disease within families suggested a genetic origin. How could a disease be simultaneously infectious and genetic? What Sir Stewart Stockman had written in 1913 was no less true in the late 1960s: "No disease is known which is both hereditary and contagious."

The second set of facts comprised the repeated failure of attempts to identify the infectious agent. Its ability to pass through the finest of filters ruled out a bacterium, and certainly a parasite or fungus. The prevailing operational definition since the turn of the twentieth century had been that any infectious agent that could pass through filters was a virus. In principle, a virus could be seen under an electron microscope, could be grown on tissue cultures, and could be purified. That is what researchers would set about doing. But to understand their work we need to understand the backdrop against which it was taking place, which had changed a great deal since the early twentieth century when viruses and genes were first defined. Since then, there had been a genuine revolution in life sciences, brought about by the advent of molecular biology. That science essentially emerged between 1950, when Vincent Zigas arrived in Papua New Guinea, and the late 1960s, when D. Carleton Gajdusek succeeded in transmitting Creutzfeldt-Jakob disease to chimpanzees.

It is said that molecular biology enabled us to learn the secret of life. Well then, what is it? What is a living being? It *exists,* it *does,* and it *reproduces.* It exists; it is a material being composed of atoms and molecules. It does; it carries out a certain number of functions. And it reproduces, either asexually—like a bacterium that grows and divides in two, giving rise to two "offspring" identical to itself—or sexually. To know the secret of life is to understand what makes it possible for a living being to exist, to do, and to reproduce.

The door was opened a crack by Mendel in the mid-nineteenth century, and a little further in the early twentieth century by the founders of genetics. It was now known that the secret of life lay in the genes. These genes had two key abilities: to tell each cell what to be and what to do, and to reproduce identically in order to be transmitted from generation to generation. But genes remained nothing more than a concept; no one knew what they were made of or how they worked.

When Pasteur's microbiology was coupled with Mendel's genetics, some light would be shed on those questions.

Before we survey the key discoveries of molecular biology, though, we must briefly review the component parts of living beings. Apart from viruses (to which we shall return later), living beings are made up of cells, which are in turn made up of molecules. Certain kinds of molecules are present in all living cells, while others are specific to certain types of cells. Some of these molecules, especially those found in all cells, are relatively simple, being composed of a small number of atoms—a few dozen—while others are far more complex and are made up of hundreds or even thousands of atoms. These macromolecules are in fact polymers: aggregates of smaller molecules linked by chemical bonds. For example, starch—the main macromolecule of the material found in potatoes and cereals—is a polymer made up of dozens or even hundreds of small molecules of the sugar glucose, its molecules in turn composed of six atoms of carbon, six of oxygen, and twelve of hydrogen.

In order to live and develop, all cells feed upon nearby molecules that contain the various atoms they need, which they transform into their own components. This requires an enormous number of chemical reactions to break down food into small, easily absorbed molecules, and to use those molecules to synthesize the cells' components. Fundamentally, a cell is defined by all the chemical reactions it can carry out, and thus by the components it can synthesize. Generally speaking, these chemical reactions do not take place on their own; each requires the presence of a specific catalyst. It has been known since the late nineteenth century that enzymes are such biological catalysts. Enzymes are themselves macromolecules of a specific class: proteins.

Hence, a cell is defined by the group of enzymes it contains. This would seem to give rise to the idea that if genes define a cell's characteristics, they must in one way or another control the formation of enzymes.

That seems unmistakable in retrospect, but it was not so obvious in the early 1940s, when it was first clearly set out by the American scientists George Beadle and Edward Tatum in what came to be called the theory of one gene, one enzyme. As we have seen, genes were initially defined as the medium for the transmission of inherited traits such as seed color in peas. Such traits were subject to natural variation: When a farmer grew a variety of peas that had seeds of a certain color, a plant would occasionally appear with seeds of a different color. Such changes were called mutations. The existence of genes was discerned by the fact that they could be the site of mutations that led to the modification of a trait. That existence was somewhat amorphous, but the experiments of Beadle and Tatum would give it a bit more substance.

There were two reasons for their success. First of all, they opted for a one-celled organism—a microscopic fungus—that would develop far faster than Mendel's peas. Second, they focused on traits far more easily defined in chemical terms than the color of a pea. These traits related to the microorganism's ability to synthesize most of the small molecules it needed to grow. Beadle and Tatum produced mutant fungi that were incapable of synthesizing one or another of those molecules; the missing molecule had to be supplied in the culture medium in order for the fungus to grow. They knew that in every instance the loss of this ability to synthesize, through mutation, resulted from an inability to carry out a chemical reaction owing to the absence of a specific enzyme. Each gene, as determined by the fact that it could be inactivated by mutation, was thus responsible for the formation of one enzyme.

So, these genes defined the enzymatic content of a cell. But what were they made of? The answer would be found in two parts, both of them during, or near, the 1940s.

The trail was blazed by the microbiologist Fred Griffith, following in the footsteps of Pasteur. He focused on a bacterium discovered by Pasteur, the pneumococcus, which could cause very serious pneumonia. With a view to developing a vaccine against that bacterium, Griffith prepared a weakened version, as Pasteur had done for the anthrax bacillus. One day in 1932, with a view to improving the vaccine's effectiveness, he inoculated a mouse not only with the weakened strain of pneumococcus, but also with the virulent form of the bacterium, which had been killed by heat. To his great surprise, the mouse soon died of pneumonia. That was strange, because neither the weakened bacteria nor the virulent bacteria that had been killed were capable of causing the disease on their own. After the necessary checking, Griffith came to the astonishing conclusion that the dead virulent bacteria had added something to the weakened bacteria, something that made them virulent. That "something" did nothing less than change an inherited trait of the weakened bacterium; he called it a "transforming factor." It seemed that the weakened bacterium had gained a gene for virulence, which it had lacked, and that this gene was the transforming factor. If a pure form of the transforming factor could be isolated from an extract of virulent bacteria, perhaps it would be possible to learn what genes were made of.

That was the bet that Oswald Avery and his colleagues at New York's Rockefeller Institute would make—and win. They published their results in 1944. Their remarkably rigorous and modest article would be one of the most important in the history of science: It showed that the transforming factor was inseparable from a macromolecular compound called deoxyribonucleic acid, or DNA. Interestingly, it had long been known that such polymers were to be found in the nuclei of cells—which were also the site of the chromosomes that were believed to carry the genes. But until then, no one had thought that this macro-molecule played a particularly important role. It was known to be a polymer containing not just one small molecule, as starch does, but four different small molecules all known as nucleic acid bases, or more sim-ply "bases." Individually, these bases are called adenine, thymine, gua-nine, and cytosine, abbreviated as A, T, G, and C.

Genes, then, are made of DNA.

From then on, this macromolecule would be a matter of great inter-est to the scientific community. It was of particular interest to two some-what atypical researchers with very different personalities, working at the Cavendish Laboratory in Cambridge, England, which specialized in using X-ray diffraction to study molecular structure. These researchers were James Watson and Francis Crick. The story of their discoveries has been told and retold, notably by Watson himself in his book *The Double Helix*.[1]

And it was indeed a double helix that was suggested for the structure of DNA. But if that had been all—setting aside its pleasing aesthetics—it would not have cast much light on how genes functioned. The impor-tant thing was the internal structure of the double helix. According to Watson and Crick, the DNA molecule was like a spiral staircase whose steps were pairs of bases. Each base was attached by means of a solid chemical bond (known as a covalent bond) to one of the "banisters" and was also linked by weak chemical bonds (known as hydrogen bonds) to a base attached to the other "banister." A fundamental principle is that a given base can be linked by hydrogen bonds to only one other kind of base: A can be linked with T (and vice versa), and G with C (and vice versa). The steps of the staircase are all AT, TA, GC, or CG steps. Those are the only possible combinations because of the structure of these bases. Watson and Crick were aware of the vast implications of the structure they were suggesting, as is clear from the final sentence of the one-page article they published in April 1953 in the journal *Nature*, one of the most famous understatements in the entire scientific litera-ture: "It has not escaped our notice that the specific pairing we have pos-

tulated immediately suggests a possible copying mechanism for the genetic material."[2]

Beneath that typically British understatement (even though Watson
was American) lie two crucial ideas that would lay the foundations of
molecular biology. The first is that, if the double helix were opened by
separating the bases of each pair—as one might open a zipper—each of
the resulting strands would be capable of defining the order in which
the free bases in the cell would polymerize with those opposite: A, T, G,
and C bases would consistently be positioned opposite the T, A, C, and
G bases of the opposite strand. The result would be the formation of
two double helices identical to the original. In other words, the genetic
material would be replicated.

The second key concept—not explicitly stated, but underlying the
article—was that the order of the four possible base pairs (AT, TA, GC,
and CG) in the double helix could constitute a code containing the information needed to synthesize proteins.

Subsequently, both of those predictions would be amply validated.

The code has been broken. Each "letter" is a sequence of three base
pairs, known as a "triplet;" there are sixty-four possible triplets. The
order of triplets in the DNA determines the order in which the constituent elements of proteins—amino acids, of which there are twenty
kinds—will polymerize.

The mechanisms that make it possible to read the code have been
identified. They involve two stages. The first is transcription, in which a
temporary unwinding of the double helix takes place, along with the
synthesis of a number of identical copies of the gene in the form of a
macromolecule very similar to DNA but composed of a single strand
rather than a double helix; this is known as messenger RNA. The second stage is translation. This involves "adapters," which recognize
triplets and their corresponding amino acids. These make it possible for
the amino acids to be lined up in the order dictated by the order of the
triplets in the messenger RNA. The translation stage takes place in
highly complex macromolecular structures known as ribosomes.

We know, then, that proteins are synthesized as linear polymers of amino acids identified by their characteristics and order. But they do not remain in the form of a more or less rectilinear thread floating within the cell. The thread folds over itself—spools up, so to speak—to form a compact structure with a perfectly defined geometry. In that form, the protein performs the biological actions that, if it is an enzyme, enable it to act as a catalyst in a specific chemical reaction.

So a gene is not just a pearl in a necklace; it is a segment of a double helix in which the sequence of base pairs determines the sequence of the amino acids in a protein. Mutations involve changes in the sequence of the bases, sometimes the simple replacement of one base with another, or the elimination or addition of several consecutive bases, entailing a change in the sequence of amino acids in the protein that the gene encodes. If that protein is an enzyme, its catalytic powers could thus be lost or altered.

In all living cells, genes are made of DNA, which is carried by the chromosomes. As we have seen, cells use a very complex process to decode the information contained in those genes. That is true for multicellular organisms such as animals and plants as well as microscopic single-celled organisms such as bacteria.

But viruses are different. They are unusual in that they are total parasites. In simple terms, they are made up of a nucleic acid protected from the outside environment by an envelope containing proteins specific to the given virus. They have genes, but because they do not have the necessary machinery to decode the information they contain, they inject those genes into a cell. The cell decodes the virus's genes as though they were its own and synthesizes all the ingredients needed for the virus to multiply, including the proteins specific to its envelope.

That, in brief, is what researchers knew in the late 1960s when they began to hunt for the infectious agent responsible for The Disease. By a process of elimination, it could only be a virus; and if that were true, it had to contain a nucleic acid.

12

THE PHANTOM VIRUS

IN FEBRUARY 1966 Gajdusek published his findings on the transmission of kuru to chimpanzees. As William Hadlow had predicted (see Chapter 10), kuru thus proved to be similar to scrapie. The same month saw the publication of an article that would puzzle the scientific community. We will address that in a moment, but first: What was known at that point about the "virus" or "viruses" responsible for these diseases? Let us focus on the one that causes scrapie, about which more complete information was available. In many respects, the scrapie "virus" was already known to be highly unusual—in its failure to trigger an immune system reaction; in its characteristic behavior during attempts to purify it; and, most important, in its extremely powerful resistance to a variety of physical and chemical agents.

When Bertrand, Lucam, and Carré observed in 1937 that no thermal reaction took place as the disease ran its course (see Chapter 1), they mentioned something that continued to baffle veterinary scientists and that would also baffle medical researchers studying kuru and Creutzfeldt-Jakob disease. This was the indication that the infected host did not marshal the defense mechanisms that would normally be used to eliminate a foreign body. In other words, its immune system was not mobilized. The absence of an inflammatory reaction in kuru patients was another sign that no immune system reaction was taking place; this made Zigas and Gajdusek, for a moment, abandon their theory that the disease was of viral origin. A more direct way to detect an immune system reaction was to look in the infected host for antibodies against the infectious agent. Antibodies are proteins synthesized when a foreign body—for example, a bacterium, a virus, or simply a macromolecule—enters the body of a vertebrate. Specific antibodies have a particular affinity for the intruder in question; they attach themselves to it, which can inactivate it, and carry it to specialized cells that will destroy it. Antibodies are found in the blood, and their presence can be detected by a number of means. But no trace could be found of an antibody specific to scrapie, kuru, or Creutzfeldt-Jakob disease. The strategy used by some infectious agents is to stop the immune system from functioning. But that theory did not work for scrapie, because affected animals showed a normal immune reaction to any other foreign molecule or particle that was injected into them. The immune system was working, but it could not detect the scrapie "virus."

The Disease had taken on the guise of an intangible thing, a phantom. It slipped into the well-defended fortress of the body unseen by the sentinels assigned to protect it.

The absence of an immune reaction also deprived researchers of a powerful means of investigation, because antibodies are extremely specific, easy-to-use tools for diagnosis and for detecting infectious agents. Without that tool, the only way to make a quantitative assessment of the "virus" was through its biological effects, its capacity to cause the disease in an animal. Even though by 1966 mice had replaced sheep and goats, such an assessment remained a long, imprecise, and costly process. That, however, was not enough to stop some purposeful scientists from trying to purify the infectious agent in the hope of being able to study its composition and its structure, and if possible to see it with the electron microscope.

But yet again, unfortunately, they met with disappointment. Separation techniques took advantage of differences in the size, mass, weight, and solubility of the various components of a cell. Whichever of those criteria was used, a given component would, in principle, be found in a specific fraction, depending on its characteristics. But the scrapie agent tended to be spread throughout all fractions. The prevailing idea was that it was associated with membrane components. Membranes are composed of fatty matter (lipids), proteins, and sugars; they form the cell walls. They are found also within cells as barriers between different intracellular structures, such as the nucleus. When cells are destroyed to prepare extracts, these membranes break up into components of various sizes; because they are fatty in nature, those components tend to "stick" to everything. So, no one had been able satisfactorily to purify the scrapie "virus" using the technology of the day.

Since it could not be purified, perhaps scientists could get an idea of its composition by finding out what it was sensitive to. Here, as described in Chapter 6, W. S. Gordon had made the first discoveries in the late 1930s, to his cost. He had observed that when treated with formaldehyde a vaccine against louping ill would transmit scrapie to vaccinated sheep. Unlike any known virus, the causative agent of scrapie was resistant to formaldehyde. In the early 1950s David R. Wilson, a veterinary scientist at the Moredun Research Institute, showed that the "virus" was astonishingly insensitive to a whole range of physical and chemical agents. Wilson, who is relatively unknown because he published very little, played an important role in the study of scrapie. Among other contributions, he showed that the "virus" was resistant to drying and to high temperatures; it was not destroyed when kept at 100 degrees Centigrade (212 degrees Fahrenheit) for thirty minutes.

The scrapie agent was a strange "virus" indeed.

Then, in February 1966, the article that would baffle the experts was published. Written by Tikvah Alper and her colleagues, it deepened the mystery surrounding the nature of the scrapie agent. At the outset, the authors used an "antimolecule gun": an electron accelerator. When fired with sufficient energy at biological matter, electrons will destroy all chemical bonds, more or less indiscriminately. The larger the target, the smaller the quantity of electrons needed to destroy it. (A whole infantry regiment firing their rifles simultaneously in the same direction stand a better chance of hitting an elephant than a housefly.) But conversely, by knowing the quantity of electrons needed to destroy a given biological activity, one can get an idea of the size of the structure in which that activity takes place. The validity of that approach had been verified by 1966—for example, by using viruses whose size had been independently determined by other methods. When applied to the scrapie agent, this technique led to the conclusion that it was about the size one would expect for a protein—in other words, far smaller than any known virus. Remember that viruses are made up of a nucleic acid surrounded by an envelope containing several proteins and, sometimes, other compounds such as lipids and sugars.

In another experiment, Alper and her colleagues used a "gun" that fired not electrons but photons. They subjected their samples to ultraviolet radiation, using a wave frequency that was absorbed specifically by nucleic acids. In large doses, this ultraviolet radiation would destroy such nucleic acids; this is the basis of a common method of sterilization that inactivates the genetic material of all microbes. But the researchers were surprised to find that the scrapie agent was completely resistant to doses of radiation forty times higher than doses that are 99 percent effective against the smallest known viruses. Here is their conclusion: "[The] evidence that no inactivation results from exposure to a huge dose of ultraviolet light, of wavelength specifically absorbed by nucleic acids, suggests that the [scrapie] agent may be able to increase in quantity [multiply] without itself containing nucleic acid."[1] Biologists were amazed and incredulous.

But then, in an article titled "Does the Agent of Scrapie Replicate without Nucleic Acid?" the same authors verified their findings.[2] The scrapie agent was, to use their word, "transparent" to ultraviolet rays— at least those of the wavelength they used, the one that destroyed nucleic acids. Bombarding the scrapie agent with ultraviolet rays was like shooting at a ghost. This called into question the central doctrine of molecular biology, that nucleic acids were the only possible vehicle of heredity, and it further highlighted the inadequacy of Pasteurian steriliization methods for infectious agents like the scrapie agent.

A period of speculation was under way. Many—the great majority— would not be convinced by the observations of Alper and her colleagues. This was not surprising, for the scientific literature is riddled with mistakes and erroneous interpretations; plus, it is always hard to renounce a credo. But some scientists began to wonder whether molecules or particles without nucleic acids could indeed be infectious. Among them was a mathematician who came to the aid of the biologists: John Griffith (not to be confused with Fred Griffith, who had discovered the transforming factor discussed in Chapter 11). In 1967, John Griffith described two mechanisms by which a protein could be infectious. In an obvious reference to the dismay shown by contemporary biologists, he started by saying that there was "no reason to fear that the existence of a protein agent would cause the whole theoretical structure of molecular biology to come tumbling down."[3]

One of the mechanisms that Griffith suggested was based on the work of the Pasteurians, a few years earlier, on the regulation of gene expression. François Jacob and Jacques Monod had started with the observation that, in bacteria, some genes are decoded—or "expressed"— only under certain conditions. They then showed that in addition to enzyme-encoding genes, there were also "regulator" genes whose job was to control the expression of other genes.

So according to Griffith's first theory, the scrapie agent could be a protein with two functions: a toxic function responsible for the disease; and another function responsible directly or indirectly for its own synthesis. In other words, that protein's gene would be present in the body's cells but it would normally remain quiescent, perhaps because a "repressor" produced by a regulator gene was preventing it from being expressed. If as the result of an infection the protein penetrated a cell, it could countermand the repressor's action so that the gene could be expressed and so that large quantities of the protein could be synthesized.

Clearly, a protein with such properties would, to recall the words of Alper and her colleagues, "increase in quantity without itself containing nucleic acid."

The other mechanism that Griffith suggested involved the idea that, once "spooled," proteins were often associated with related proteins, either in pairs to form "dimers" or in larger numbers to form trimers, tetramers, and so forth. Griffith's idea was that potentially infectious protein existed within cells but in a nonassociated form, as a nontoxic monomer. This would be incapable on its own of giving rise to a toxic dimer. But if it were in the presence of a toxic dimer that had penetrated the cell, it could associate with the dimer to form a trimer and then a tetramer, which could in turn divide into two dimers. One toxic dimer would thus result in two toxic dimers, which would by the same mechanism become four, then eight, and so forth. By this process, a nontoxic form of a protein could become toxic through contact with the toxic form. That theory was to be prophetic.

But irrespective of Griffith's speculations, the nature of the causative agent of scrapie and related human diseases remained a complete mystery as the 1960s came to a close. This would not change over the following ten years, during which a tragedy would unfold in a completely different area of medicine. An amazing advance in pediatric endocrinology was to pave the way for a counterattack by The Disease. To understand what happened, we must return to 1959.

13

A TRAGEDY IN THE MAKING

HERE IS WHERE THE HUNT for The Disease stood in 1959: British veterinary researchers had transmitted scrapie to goats, but not yet to mice. The writings of Creutzfeldt and Jakob were gathering dust on library shelves. Gajdusek had left the Fore people and had learned of Hadlow's theory positing a parallel between scrapie and kuru. The year 1959 also saw the early triumphs of molecular biology. Watson and Crick had published their historic article six years before, and Jacob and Monod would publish their initial work on the regulation of gene expression.

That was the context in which certain types of dwarfism began to be treated with growth hormone extracted from human pituitary glands.

Hormones are chemical messengers that play an essential role in co-ordinating the development and functioning of multicellular organisms, specifically humans and animals. They are secreted by endocrine glands,[1] and their chemical nature is extremely diverse. For example, insulin is a protein whereas sex hormones are steroids (complex molecules related to cholesterol). Once they have entered the bloodstream, hormones find the target cells they are intended to activate (*hormone* is from the Greek for "to set in motion") because these cells contain specific receptors, which are generally proteins. Interaction between a hormone and its receptor sets in motion a biochemical process that causes the target cell to carry out its proper function.

Insulin is a well-known example of such a hormone. It is secreted by cells in the pancreas, and is a small protein made up of about fifty amino acids. Its purpose is to reduce glucose levels in the blood. To do this, it acts on cells of many kinds, stimulating their ability to capture and break down that sugar. Insulin deficiency causes a serious disease—insulin-dependent diabetes mellitus—which is fatal if not treated. An effective treatment was developed in 1921 using injections of insulin from animal sources.

The successful treatment of diabetes with insulin was undoubtedly encouraging for researchers seeking a treatment for pituitary dwarfism. The pituitary, or hypophysis, is a small endocrine gland located at the base of the brain and connected to it by a thin stalk. It provides a key link between the nervous and endocrine systems and is sometimes described as the conductor of the endocrine orchestra. It produces a number of hormones that control the activity of other endocrine glands, including thyrotropin—thyroid-stimulating hormone, or TSH—and adrenocorticotropic hormone, or ACTH, which regulates the production of hormones by the adrenal glands.

The role of the pituitary in the body's growth had been demonstrated in 1916, when experiments showed that removal of the pituitary from tadpoles halted their growth, and that this could be corrected by injecting pituitary extracts. A few years later, in 1921, other researchers showed that injecting cow pituitary extracts into rats caused gigantism and that removal of the pituitary from dogs would halt their growth. It seemed reasonable to conclude that a pituitary hormone stimulated growth. Scientists purified the hormone using the pituitaries of various animals, testing the ability of the different extracts to stimulate growth in rats whose pituitaries had been removed. This growth hormone, known also as somatotropin, turned out to be a protein made up of a chain of 191 amino acids.

Human growth depends on many genetic and environmental factors. For example, there is a certain correlation between the height of parents and that of their children, and factors such as diet also play a very important role. In principle, growth takes place in a balanced and predictable way. In the developed world, where children are regularly examined by pediatricians, one of the first things a doctor does on each visit is to measure the child to ensure that growth is normal. Any significant deviation from the normal growth curve can be a sign of a disorder. And in some cases, slow growth turns out to be due to a growth hormone deficiency.

Such deficiencies can be either total or partial. They can result from pituitary tumors or from head injuries, but most often they are of unknown origin. Doctors mask their ignorance with scientific jargon: They call such a disease "idiopathic." Growth hormone deficiency in children, especially when total, can result in the very serious condition known as pituitary dwarfism, in which height at adulthood does not exceed four or four and a half feet, and which also entails metabolic problems and, sometimes, malformation. This is serious both physically and because of its psychosocial effects.

Once growth hormone had been identified and purified, as it had been by the early 1950s, there was a great temptation to try it as a treatment for pituitary dwarfism, on the model of using insulin to treat insulin-dependent diabetes. Here, medical researchers were to run into a major problem. Despite small differences in the amino acid sequences of human and pig or cow insulin, the latter worked in humans and were easily obtained. Growth hormone was different. Primates reacted only to growth hormone derived from primates. Monkeys with their pituitaries removed did not react to injections of growth hormone from swine or cows. So treating pituitary dwarfism with growth hormone demanded that the hormone be extracted from human or simian pituitaries. It would not be feasible to obtain monkeys in large enough numbers, so the only possible source was humans. It would be necessary to remove the glands from recently deceased people, as was already being done for other organs such as corneas.

The first attempts were made in the late 1950s. In 1957, the American Maurice Raben described the earliest method for purifying growth hormone from human pituitaries. Because the hormone was to be used for medical treatment, he made sure that his method included steps that "provided strong bactericidal and viricidal action in the extraction of human pituitaries of indeterminate origin."[2] The following year he reported that a seventeen-year-old boy with pituitary dwarfism had been given two or three intramuscular injections of the hormone per week for ten months. This treatment was well tolerated and effective; it immediately increased growth rate fivefold. This first very encouraging outcome sparked much research that would benefit thousands of children with dwarfism. In 1959, treatments began in the United States and the United Kingdom.

An early goal of doctors and researchers was to improve purification techniques, and a number of these were devised. They increased the purity of the product by stripping it, inter alia, of other pituitary hormones; they achieved higher yield, which was important because of the difficulty of obtaining human pituitaries; and, to the extent possible, they reduced the risk of bacterial or viral contamination. Purification was a long and complex process, and obviously could not be done using individual pituitaries. So glands collected in hospitals were frozen and stored; when there were enough of them, they were combined and the extraction process was carried out on the entire lot. Depending on the laboratory, the size of the lot would range from a few hundred to a few thousand pituitaries. The figure of twenty thousand has been mentioned for some lots in the United States.

Work was done also to refine the treatment protocol and to determine its effectiveness. Here, I must cite a British study carried out by a Medical Research Council (MRC) working party, which began in the early 1960s and whose results were published in 1979. The study focused on some six hundred patients treated with hormone prepared in MRC labs between 1959 and 1976. The outcome left no room for doubt about the effectiveness of the therapy, although this depended on a number of factors such as the cause and nature of the hormone deficiency, the age at which treatment began, and the duration of the treatment. Hormone therapy was thus an undeniable success. Still, the authors were worried, writing: "If, as is hoped, patients are diagnosed younger and more patients with partial deficiency are recognized, demand may soon outstrip supply."[3] That was the price of success. This concern was to be a leitmotif for hormone producers, both academic institutions such as the MRC and pharmaceutical companies involved in the hormone market. In the United Kingdom the number of children being treated would rise to about eight hundred. Given that some fifty pituitaries were needed to treat one child for a year, thirty thousand to forty thousand glands would be needed annually.

Meanwhile, in France a small number of pediatricians, encouraged by the success of their American and British counterparts, began to use the new treatment during the 1960s. This was in an experimental context and under difficult conditions, because the hormone was in short supply. The successful outcome convinced doctors that it would be beneficial to make the therapy available to all children who needed it. This meant that the hormone had to be produced in France in order to avoid dependence on scarce and expensive foreign products. So in 1973 L'Association France Hypophyse (the French Pituitary Association) was set up on the initiative of prominent pediatricians and the French government. Its purpose was to coordinate and organize a complex set of operations ranging from harvesting pituitaries, through purifying and distributing the hormone and preparing it for administration, to selecting priority patients for the therapy. Jacques Monod—Nobel laureate for medicine in 1965 and then director of the Institut Pasteur—willingly agreed that one of the institute's laboratories should be responsible for extraction and purification. This was in keeping with the tradition of public service that the Institut Pasteur had maintained since it was founded in 1888. At the time he took that decision, Monod appears not to have thought that such activities could pose any risk at all—least of all that of transmitting Creutzfeldt-Jakob disease.

Between 1959 and 1985, human growth hormone therapy gave complete satisfaction. There were no complications, and the number of candidates continued to increase. The limiting factor was the availability of hormone, which was a function of how many pituitaries could be harvested. By 1985, a total of approximately twenty-five thousand children had been treated worldwide. In the case of the earliest patients, it had been a quarter-century since their treatment had begun. There was universal applause for what seemed to be one of the great achievements of modern medicine.

But, unbeknownst to anyone, The Disease had struck again. The dream was to turn into a nightmare.

14

ONE CASE PER MILLION

AS A RARE, POSSIBLY GENETIC AILMENT, Creutzfeldt-Jakob disease was of
interest to only a handful of doctors and researchers between 1920 and
1960. But there had been enough interest to ensure the gathering of data
on patients who had been treated in several countries. During the 1960s,
the question of whether CJD was a single disease became a matter of increasing controversy. Citing the diversity of clinical symptoms and of
the lesions observed in nervous system tissues, some were convinced
that several different diseases were gathered under the Creutzfeldt-
Jakob umbrella, and they came up with as many as sixteen names for
these. Others stressed the common characteristics and argued that it was
indeed a single disease. It was not until the 1970s, when Gajdusek had
demonstrated that CJD was transmissible, that the latter viewpoint
began to prevail.

At this point, specialists summed up the symptoms and general
course of the illness as follows: CJD principally affects men and women
aged between forty and sixty and lasts about a year. It usually begins in-
sidiously with vague symptoms such as anxiety, difficulty in concentrating or speaking, memory loss, and problems with walking. After several
weeks, more pronounced symptoms appear, including paralysis in one

or more limbs, tremors, jerky involuntary movements, fits resembling epileptic seizures, and dementia (loss of higher cortical functions such as reasoning and memory). These symptoms, whose relative severity varies from patient to patient, reflect a general attack on various parts of the central nervous system, including the cerebellum and the medulla in the brain. Apart from these neurological symptoms there is no fever and there are no changes in blood chemistry or in the proteins of the blood serum or the cerebrospinal fluid. Electroencephalograms (EEGs), on the other hand, often show a characteristic pattern that can aid in diagnosis. CJD develops gradually, with neither remissions nor flare-ups. In the final stage, the patient is in wretched condition: stupor, inability to speak, varying degrees of paralysis, loss of basic functions, and then rigidity and coma. Death follows either from no apparent cause or from an infection, often pneumonia.

On autopsy, the general appearance of the brain is normal, but microscopic examination reveals that the gray matter is spongy owing to the presence of vacuoles both within the brain cells—neurons—and in the extracellular matter. Those are the loci of degeneration that Creutzfeldt and Jakob had observed. Some kinds of cells other than neurons—astrocytes in particular—appear overdeveloped in size and sometimes in number. This microscopic examination of brain tissue is essential to confirm the diagnosis; it is usually conducted as part of the autopsy because biopsies, which can exacerbate the patient's condition, are generally avoided.

Because of the uncertainty even about diagnosis, no serious epidemiological study was begun until the early 1970s. But one question was clarified around that time: Creutzfeldt-Jakob disease did not *systematically* appear in families. Further cases within families had been described, but in general no cases could be detected either among patients' forebears or among their descendents. The term for this was "sporadic cases."

Then, the first results of epidemiological studies began to appear in the 1970s. At the outset, a pair of independent incidents drew the attention of researchers: two likely instances of human-to-human transmission following surgery. The first involved a fifty-five-year-old American woman who had received a cornea transplant. The cornea came from a man who had just died of pneumonia. Eighteen months later, she experienced the initial symptoms of what would be diagnosed as Creutzfeldt-Jakob disease, of which she would later die. It turned out that the cornea donor too had had CJD; he showed some of its symptoms, but the final diagnosis could not be made until the autopsy—which took place after his cornea had already been removed and transplanted. A brief report in 1974 stated that it was very probable that the disease had been transmitted with the transplant.

The second case was reported in 1977. It concerned two young patients, a twenty-three-year-old woman and a seventeen-year-old boy, both of whom initially suffered from epilepsy. Their treatment involved the taking of EEGs, which required the insertion of electrodes into the brain. Two and a half years after the procedure, both showed the initial symptoms of Creutzfeldt-Jakob disease. Because they were unusually young, doctors looked for possible sources of contamination. It appeared that two of the electrodes used during their EEGs had been used a few weeks earlier on a sixty-nine-year-old woman with CJD. They had, of course, been sterilized as usual with alcohol and formaldehyde. But, as we know, this would have had little effect on the CJD agent. Here again, the conclusion was probable transmission of that agent.

Following these probable cases of transmission—which were declared to be iatrogenic, which means they were caused by a medical procedure—neuropathologists and surgeons were told to take precautions in handling biological matter from both CJD patients and possible CJD patients.

In 1973 a group of scientists studying CJD epidemiology in Israel made a surprising observation. Although the average prevalence was about one case per million inhabitants, it was thirty times as high among Jews of Libyan origin. What could be the reason for this amazing rate of incidence? Gajdusek and his colleagues suggested that they could have been contaminated by the infectious agent of scrapie by eating grilled sheep's eyes, a delicacy for certain groups in North Africa, including Libya. That theory depended on the presence of the scrapie agent in the eye. As we have seen, a case of CJD transmission through a corneal transplant had just been reported. Initially, the researchers were skeptical because Libyan Jews are not the only people who eat sheep's eyes or, especially, brains, but eventually they were convinced—sufficiently so to suggest, on the basis of some additional observations made outside Israel, that in general Creutzfeldt-Jakob disease could be caused by the consumption of animals infected with scrapie.

To some degree at least, this notion was prophetic: Twenty years later, transmission of The Disease from cows to humans, through food, would be very much in the headlines.

Yet the observations behind this idea were to be given another interpretation that had nothing to do with eating sheep's eyes or brains. This emerged in 1979, with the first indications that the high incidence of CJD among Libyan Jews living in Israel could have a genetic explanation. We shall be returning to this point later.

The first epidemiological study of any scale was published in 1979 by a group of researchers including Gajdusek. Based on analysis of the medical records of 1,436 patients, it drew the following conclusions that for the most part remain valid today:

· Creutzfeldt-Jakob disease existed in every country for which data were available.

· In each such country, annual mortality from CJD was between 0.5 and 1 per million inhabitants.

· Geographic distribution of cases was random, with a few exceptions (five at that time) in which cases were concentrated in a particular area, as with the Libyan Jews living in Israel.

· Some 15 percent of cases were in families in which at least two members were or had been affected; these were described as familial.

The authors of this study then addressed the possible origin of the disease. They considered both human-to-human contagion and contamination through food by way of the alimentary tract (which assumed that CJD could be a human adaptation of scrapie).

As to human-to-human contagion, they found that a certain percentage of patients had indeed undergone surgery or were engaged in high-risk professions such as medicine or dentistry. But no convincing conclusions could be drawn from these data. The small number and the geographic distribution of cases made it most unlikely that contamination could have taken place through direct contact between individuals. Moreover, the authors noted that no case of CJD or kuru had been observed among the many people who had lived in the area where kuru flourished but who had not practiced cannibalism.

Furthermore, they felt the theory that scrapie could be transmitted to humans through food was very unlikely. No correlation had been found between the incidence of CJD in humans and scrapie in sheep within a given area. For example, the incidence of CJD was the same in Australia, where no cases of scrapie had been reported for twenty years, as it was in the United States, where scrapie had been present for many years. Returning to the Libyan Jews, the authors noted that scrapie seemed not to exist among North African sheep at all. The theory of contamination through consumption of the eyes or brains of infected animals thus seemed even less likely.

They closed their article with this comment: "We must conclude that the natural mechanism of spread and the reservoir of the CJD virus remain unknown at present."[1] The authors remained convinced that Creutzfeldt-Jakob disease was caused by contamination from the "virus" of that disease—but admitted that they were unable to identify its source or the way in which it was transmitted.

15

PRIONS

DESPITE THE AVALANCHE OF THEORIES loosed by the observations of Alper and her colleagues, nothing spectacular was published during the 1970s about the nature of the mysterious "virus" that was seemingly without nucleic acid.

Enter a man who continues to occupy center stage to this day: Stanley Prusiner, a neurologist and biochemist who since 1974 has divided his time between the San Francisco and Berkeley campuses of the University of California. Starting in 1978, but primarily in the early 1980s, he published an impressive series of articles that identified what some say is the causative agent of scrapie and what others view as a principal component of that agent. Undeterred by the many problems his predecessors encountered, Prusiner set about attempting to purify this mysterious "virus." Here, he was able to take advantage of considerable recent progress in purification techniques. For one thing, he had highly sophisticated instruments that had not existed ten years earlier. But that was not enough; the outcome described in his first important article on the subject, in 1978, was not particularly encouraging. Like those before him, he observed that the agent was found in nearly all fractions, no matter what separation method he used. But he did not give up.

One major problem was the duration and cost of his experiments, even when he used mice to measure the biological activity of the agent. Just imagine: To measure the quantity of the agent in a single sample he needed to prepare a series of ten or so different dilutions (serial dilutions), to inoculate six mice with each dilution, then to observe those sixty mice every week for a year to see if any scrapie symptoms had appeared. Separation techniques yielded between ten and fifty fractions, which meant that from six hundred to three thousand mice had to be inoculated and observed for a year in order to know the quantity of infectious agent in all fractions.

One improvement was to replace the mice with hamsters, in which the incubation period was shorter. Another—which might seem a step backward—was to replace the serial dilutions with a measurement of the incubation period: the greater the quantity of agent, the shorter the incubation period. That was the method used by veterinary researchers in the days when their only experimental animals were sheep and goats, but at that time they had had no way of quantifying the results. By comparing measurement by incubation period with traditional measurement by serial dilutions, Prusiner developed a new method that for a given sample required only four hamsters to be inoculated and observed for two months—four hamsters instead of sixty mice, and two months of observation instead of a year. That was a major advance.

There were biochemical problems too, raised by the very unusual behavior of the agent that he was trying to purify. But this is not the place to discuss these issues, even though resolving them proved to be quite a feat.

So what was the outcome? Well, the mysterious, ghostlike agent, purified from hamster brain tissue, was protein in nature. And in line with the predictions of Alper and her colleagues, it lacked nucleic acid. The most completely purified fraction contained a type of protein that represented 90 to 95 percent of the proteins that were present in the fraction. Given an equal quantity of protein, this purified fraction had five thousand to ten thousand times the infectious potency of the initial cellular extract. The size of this protein suggested that it was made up of a chain of about three hundred amino acids, which is in the normal range for proteins in general.

It was highly unorthodox to say that a virus was protein in nature and that it contained no nucleic acid. So Prusiner and his colleagues worked to find additional proof that this was indeed true for the scrapie "virus." They demonstrated a strict correlation between the quantity of the protein they had identified and the infectious potency of the corresponding fraction. They showed that, although exceptionally resistant to most physical and chemical agents, the infectious potency of the purified fractions could be destroyed by any of several agents known to destroy or inactivate proteins. On the other hand, potency was completely resistant to all treatments known to destroy or inactivate nucleic acids.

There was every indication that the protein Prusiner had purified was the scrapie "virus"—or at least one of its essential components. It was impossible to completely rule out the presence of a nucleic acid in this "virus," but it would have to be extremely small, containing a few bases at the very most. In 1982, Prusiner named this new kind of infectious agent. He called it a "prion": "Prions are small *pro*teinaceous *in*fectious particles which are resistant to inactivation by most procedures that modify nucleic acids. The term 'prion' underscores the requirement of a protein for infection; current knowledge does not allow exclusion of a small nucleic acid within the interior of the particle."[1] Prusiner remained cautious with respect to the possible presence of nucleic acid, but the key role assigned to protein was reflected in his choice of terminology. The word *prion* is based on "*pro*teinaceous *in*fectious particle" ("proin" would have sounded less euphonious).

Let us spend a moment on two of this protein's special characteristics: its resistance to proteases, and its strong tendency to form aggregates.

Proteases, also called proteinases, are enzymes that destroy proteins. They attack the bonds linking amino acids in a protein and hydrolyze (destroy or "cleave") them. It may seem odd that there are enzymes—which are themselves proteins—whose job it is to destroy other proteins. In fact, proteases have many functions, including a "digestive" one: to hydrolyze other proteins so that their amino acids can be reused. Protease action is kept under tight control within a cell to prevent major damage to the cell's own proteins. Proteases are often segregated in special compartments or rendered dormant by inhibitors. And not all proteins are equally sensitive to proteases. As a general rule, the more tightly spooled a protein's amino acid chain, the more protease-resistant it will be. And the prion protein proved to be particularly resistant to these enzymes. They could, of course, hydrolyze it, but there were conditions under which it was resistant while nearly any other cellular protein would be smashed to pieces.[2] This made it far more easy to purify, but it was also interesting in another way (which will be discussed in Chapter 17).

Another property of this protein is its tendency to form aggregates or polymers. The traditional way to determine the size of a protein's amino acid chain starts with treating the protein with powerful detergents in order to dissociate any polymer or aggregate it might form. That was how Prusiner concluded that the prion protein contained a long chain of some three hundred amino acids. But it is also possible to determine the size of a protein without first treating it with a detergent, and if the answer is the same as it was with detergent treatment, that is because the protein is in nonaggregate, nonpolymer form: a monomer. But that was not what researchers saw in the case of the prion protein. The purified fractions with the highest infectious potency behaved like a mixture of different-sized particles; they could be as much as a thousand times as large as a protein made up of three hundred amino acids. Electron microscope study of these preparations showed that they contained rodlike structures of various lengths, often bonded to one another. These rods were polymers containing dozens, hundreds, or even thousands of copies of the prion protein. In retrospect, the protein's tendency to aggregate would largely explain the difficulty of purifying it. As we shall see, such aggregation was not an experimental artifact.

Recall that Klatzo, Zigas, and Gajdusek, when examining brain tissue from kuru patients, had been able to recognize what they called amyloid plaques by their color when exposed to certain products. These were apparently made up of tangles of fibrous structures (fibrils); they bore a certain resemblance to images seen in other degenerative nervous system diseases, such as Alzheimer's disease. Amyloid plaques very similar to those of kuru were later noted in some cases of Creutzfeldt-Jakob disease and in the brains of various animals with scrapie. Prusiner and his colleagues initially noted that the rods in their prion protein preparation had optical and color characteristics similar to those of amyloids. And they wondered whether these rods could exist as such in the brains of sick animals, forming these amyloid plaques. This was confirmed through the use of antibodies directed against the prion protein.

And yes, Prusiner had finally succeeded in finding such antibodies. He did not find them in animals with scrapie—for, as we have seen, they do not produce antibodies—but he obtained them by injecting rabbits with relatively large quantities of prion protein purified from the brains of infected hamsters. Once you possess antibodies that are specific to a given biological molecule, many research opportunities arise. For example, you can attach "markers" to an antibody that are visible under the optical or electron microscope. If the antibody is placed on a tissue section, it will bind to its target molecule, and the location of the visible marker will indicate the location of the antibody in the tissue. Prusiner's group used that technique to show that antibodies against the monomer prion protein bound to the rod structures, which proved that these were indeed made up of prion protein. They went on to show that these rods, which were often tangles of filaments, were present in brain tissue extracts from animals with scrapie. These filaments had previously been observed by other researchers and had been described as fibrils. Amyloid plaques were the apparent result of the combining of a great many such fibrils. The fibrils and the amyloid plaques were located between nerve cells; they had probably been liberated into the extracellular space following the destruction of cells as the disease progressed.

All these findings had referred to scrapie, but Prusiner and his team would extend them to Creutzfeldt-Jakob disease. A protein with properties identical to those of the scrapie prion could be observed in the brains of CJD patients.

Thus, during the first half of the 1980s there had been a major leap forward in the hunt for The Disease. It seemed to be caused by a protein called a prion. This astonishingly stable protein accumulated in the brains of sick humans and animals in the form of filaments or fibrils that sometimes in turn combined to form amyloid plaques. But where did this protein come from, and how did it multiply? The answers would come in the second half of the decade.

16

APRIL 1985

APRIL 1985 WAS A KEY MONTH. It seemed as though The Disease had been flushed out. But it counterattacked on two fronts, launching deadly offensives whose effects would still be felt fifteen years later.

As described in Chapter 15, Prusiner believed that The Disease was caused by a kind of protein, which he named a prion. It is not hard to imagine a toxic protein, for there are many precedents. The toxins produced by the bacteria that cause diphtheria, anthrax, and botulism are proteins, and they can kill humans and animals in extremely low doses. But they do not reproduce; on postmortem examination, the body of an animal will contain no more of the toxin than it had been inoculated with. So, because the toxin is diluted in the body, samples drawn from the animal will be harmless. It is not the same with prions. As we have seen, a mouse that has died of scrapie will contain enough prion to infect millions of other mice. That is usual for viral infections; the virus multiplies in the body of the victim, whose infected organs eventually contain much larger quantities of virus than the original inoculation. But viruses contain nucleic acid, which provides the information needed for multiplication, whereas prions appeared to have no nucleic acid. So how could their ability to multiply be explained?

It was not reasonable to think that a protein could contain the information needed for self-synthesis. Only a gene—a segment of a nucleic acid—can put the amino acids that comprise a protein in the right order. A prion gene thus had to be found within the infected organism. That was the hypothesis underlying John Griffith's description of mechanisms by which a protein could be infectious. Now that a protein had been exposed as the alleged culprit, would it be possible to learn whether its gene was in fact present in the infected organism? In the late 1960s, when Griffith offered his theory, that question could not be answered, but by the mid-1980s it had nearly become child's play. By then, extremely powerful genetic engineering technologies had been developed. For example, it was possible to cut DNA at precise points, to manipulate the resulting fragments, and to insert them into the chromosomes of other organisms. Within these DNA fragments, it was possible to determine the order of the bases and thus to learn the order of the amino acids within coded proteins from the genes they contained. And conversely, it would be possible to find a protein's gene by knowing the order of a protein's amino acids. That is what the genetic engineering expert Charles Weissmann and his group did in Zurich, in collaboration with Prusiner's group and a group that specialized in identifying amino acid sequences in proteins.

The outcome was published in April 1985: The prion gene was present in hamsters whether or not there was any infection. It was possible to deduce from the sequence of bases in this gene that it encoded a protein containing 240 amino acids. A similar gene had been found in other mammals, including mice and humans. But if the gene was present in the chromosomes of these animals, why did it not bring about the synthesis of a prion protein in the absence of any infection?

John Griffith had suggested one possibility: that the gene could be expressed only in the presence of the protein that acted as a regulator. But that theory was soon ruled out. The expression level of a gene could be measured by the concentration of its primary product, messenger RNA. That level was the same in the brains of infected and uninfected animals, so the gene was transcribed the same way in both. In other words, the infectious protein did not cause its own gene to be transcribed and was not a regulatory protein as the term was understood by Griffith.

What about translation of messenger RNA—in other words, protein synthesis? To find out, the authors used ultraspecific reactants: prion protein–specific antibodies. They found that the protein was present in the brains of uninfected animals, but in smaller quantities and with properties different from those of infectious prion protein. In uninfected animals it could be destroyed by proteases, in particular proteinase K, under conditions where infectious prion protein was resistant to them. That was why Prusiner had not detected it in brain extracts from uninfected animals; in order to purify the prion, he had treated the extracts with proteinase K.

Thus, the prion gene was present in brain cells, but the structure of the protein that encoded it differed depending on whether or not the animal was infected. This structural difference meant a difference in protease sensitivity.

As usual in science, every answer raised new questions. Here, the most obvious ones related to the exact nature of the differences between the infectious protein and the protein normally synthesized in the brain, and, most important, to the mechanism by which a single gene could control the synthesis of two kinds of protein, one protease-sensitive and noninfectious and the other protease-resistant and infectious. But for the present, the researchers had every reason for satisfaction. The scrapie agent had become far less mysterious.

As though The Disease sensed their satisfaction, however, it chose that moment to demonstrate its deadly power. For it was in April 1985 that we learned the growth hormone used to treat pituitary dwarfism could carry the agent that caused Creutzfeldt-Jakob disease. What exactly took place that month?

The tale is dramatically told by a former colleague of Gajdusek, the American scientist Paul Brown, himself one of the greatest living experts on spongiform encephalopathies:

In May 1984, [a] young man and his family flew from San Francisco to Atlanta en route to Maine to visit his grandparents. As he rose from his seat to change planes, he complained of dizziness. His mother, who was experienced in the diagnosis of hypoglycemia, gave him some candy and watched him closely for the rest of the trip. Nothing more happened, and the incident was forgotten. Several days later, however, in Maine, he turned down an offer to go for a spin on the lake in his grandfather's motorboat, saying that "he didn't need to go for a spin because he was already dizzy." On his return from Maine, the patient went back to school, but his dizziness persisted, and now his speech seemed slightly changed.[1]

At this point, the family sought medical advice. In June, the young man went to see the pediatric endocrinologist Dr. Raymond Hintz. The family knew him well, because the young man had a long history of hormone treatments. In 1965, at age two, he was diagnosed with deficiencies in thyroid hormone, insulin, and growth hormone. For fourteen years beginning in 1966, he received injections of human growth hormone, which enabled him to reach a reasonable height. Hintz referred his patient to neurologists. His condition grew worse, with symptoms of nervous system deterioration, but the specialists could not agree on a diagnosis. Creutzfeldt-Jakob disease was briefly considered, but was quickly rejected because of the patient's youth. But after his death that November, an examination of his brain showed that this diagnosis had been correct.

When he was informed of the result a few months later, Hintz might have wondered whether this had any relation to the occasional cases of iatrogenic transmission of CJD that had been reported in the medical literature. In any event, on March 4, 1985, Hintz wrote to the Food and Drug Administration (FDA), which is responsible for approving drugs for sale in the United States. In his article, Brown quotes from that letter: "The patient was treated for 14 years with growth hormone, and I feel that the possibility that this was a factor in his getting Creutzfeldt-Jakob disease should be considered. A careful follow-up of all patients treated with pituitary growth hormone in the past 25 years should be carried out, looking for any other cases of degenerative neurological disease."[2]

The FDA did not need to be told twice. It immediately notified doctors about the problem and asked them to report any similar cases. Responses came quickly: On April 11, a Dallas doctor reported that one of his former patients, who had been treated for a pituitary deficiency, had died in February 1985, at age thirty-two, of an unidentified neurological disease. A doctor from Buffalo reported on April 18 that a twenty-three-year-old man who had also been treated with human growth hormone had likewise died of an unidentified neurological disease. In the latter case, a subsequent diagnosis of Creutzfeldt-Jakob disease had been made following examination of the brain.

On April 19, 1985, the FDA suspended approval of treatment using human growth hormone extracted from pituitaries. It was indeed hard to imagine that it was mere coincidence that three certain or probable cases of CJD had been reported among the approximately ten thousand young Americans treated with human growth hormone. Not only is CJD extremely rare—about one case per million individuals per year—but it is even rarer in young people—approximately one case per hundred million persons under age forty. The most likely interpretation was that one or more pituitaries from people who had died of CJD had been included in the lots used to prepare the hormone, and that the purification process had not been sufficient to remove the infectious agent. American authorities were even less hesitant about their decision because a human growth hormone prepared differently—through genetic engineering—was soon to be marketed.

Soon afterward, another case was reported, this time in the United Kingdom. In February 1985, a twenty-three-year-old woman who in childhood had been treated with human growth hormone died of Creutzfeldt-Jakob disease. That made four cases in the space of a few months. Were we witnessing the start of an epidemic that could affect many of the approximately twenty-five thousand patients in several countries who had been given human growth hormone treatment in childhood? Or were these isolated cases in patients who had been given hormone prepared using an old-fashioned method and inadequately purified? The truth would lie somewhere in between.

In the meantime, while doctors were asking these questions, English farmers found a strange new disease affecting their cattle. The first case was observed in April 1985, but others soon followed. The symptoms were described in this way:

> Previously healthy cattle, in good bodily condition, became apprehensive, hyperaesthetic and developed mild incoordination of gait. Their mental status was progressively altered and normal handling procedures evoked kicking. Fear and aggressive behaviour were recorded and auditory stimuli produced exaggerated responses, even falling. The incoordination of gait gradually became more pronounced with hypermetria and falling. Eventually frenzied behaviour and unpredictability in handling, or recumbency, necessitated slaughter. Clinical pathology did not support the diagnosis of any hitherto recognised diseases of cattle.[3]

The reader will have surmised that these were the early symptoms of so-called mad cow disease, the latest avatar of The Disease. But it was only later, toward the end of 1986, that, having examined the brains of sick animals, veterinarians came to the same conclusion and "provisionally" named the disease bovine spongiform encephalopathy, or BSE.

So April 1985 saw the early signs of two grave public health crises that would be particularly severe in Europe, especially in France for one (contamination of human growth hormone with the infectious agent of CJD) and in the United Kingdom for the other (mad cow disease and its transmission to humans). On the other hand, the same month had also seen a dramatic advance toward identifying the cause of The Disease: proof that the prion gene was present in the cells of all mammals.

17

THE "KISS OF DEATH"

LET US LEAVE THE DOCTORS and veterinarians to their confusion and concern, and return for a moment to the guilty party—the causative agent of The Disease, the prion. As we have seen, this appeared to be a protein encoded by a gene that was present, with very similar sequences, in all mammals. Normally, this gene would encode a protein containing 240 amino acids that, like most other proteins, was protease-sensitive. But in animals with scrapie or humans with Creutzfeldt-Jakob disease, the protein would take a different form, one highly resistant to proteases. This other form was infectious; inoculating animals with it would cause the accumulation of a protein with the same properties. We thus speak of "normal" and "infectious" proteins to distinguish the two forms.

In the hope that they could learn the secret of its unusual properties, scientists closely studied the protein in both its normal and its infectious forms. First of all, in 1986, they learned that uninfected animals synthesized only the normal protein, while infected animals synthesized the infectious protein plus the usual quantities of the normal protein. The normal protein proved to be a membrane protein, located on cell surfaces. It contained a lipid attached to one end of the chain, which anchored the protein to the membrane. The infectious protein contained the same lipid, but it was not anchored to the surface of the cell. It was probably bonded to intracellular membranes.

Another important difference between the normal and infectious forms is the speed at which they are synthesized and degraded. In an infected cell, the infectious protein is synthesized far more slowly than the normal form. It appears that the infectious protein is a secondary formation on the basis of the normal protein. Proteins, once synthesized, do not last forever. They are periodically replaced at a rate that varies from protein to protein. This involves a period of degradation caused by proteases. Here, the normal form of the prion protein behaves like other proteins; it is degraded relatively quickly. But because it is protease-resistant, the infectious protein is highly stable; no degradation has been observed under experimental conditions. So the normal protein is periodically replaced and remains at a more or less constant concentration, while the infectious protein accumulates, reaching high concentrations. This accumulation may be the reason for cell destruction and thus for the gradual appearance of nervous system lesions.

The secret of some or all of the differences between the normal and the infectious protein probably lies in the way they fold. As we have seen, proteins are long threads—chains of amino acids that are folded, or spooled, resulting in a compact structure with a well-defined geometry. A number of physics-based analytical techniques enable us to know this geometry. These include X-ray diffraction, which requires the use of protein crystals (which can be difficult to obtain), and nuclear magnetic resonance (NMR), which is carried out on the protein in solution. Despite vast progress over the past two decades, identifying the structure of proteins by these techniques remains difficult, or sometimes impossible, owing to the properties of a given protein. Today we know the structure of a considerable number of proteins—more than ten thousand, in fact. Study of these structures has yielded some rules about the way in which amino acid chains are folded within proteins. The various segments of the chain are basically in one of three configurations: helical (alpha-helices); in zig-zags on a flat plane (beta-sheets); and structures that, unlike alpha-helices and beta-sheets, are irregular and are generally located in between areas with a regular structure and at the ends of the chain. Many weak bonds among the various segments of the chain give the protein its compact structure.

But what about the prion protein? In 1993 Fred Cohen, together with Prusiner and other colleagues at the University of California at San Francisco, discovered a structural difference between the normal and the infectious protein. Techniques less complicated than X-ray diffraction and NMR could provide an idea of the alpha-helices and beta-sheets contained in a protein. Using these techniques on the prion protein showed that the normal form was rich in alpha-helices and had few or no beta-sheets, while the infectious protein was rich in beta-sheets. On the basis of what they had learned about the prion protein and about the structures of other proteins, Cohen, Prusiner, and their colleagues went on to propose hypothetical structures for both the normal and the infectious forms of the prion protein. They suggested that the normal form contained four alpha-helices, and that the infectious form contained two alpha-helices and four beta-sheets. The switch from one form to the other that was thought to occur during the course of the infection involved a fairly major structural modification leading to the molecule being "compacted." In spite of their considerable aesthetic appeal, however, these hypothetical structures would soon be called into question.

The true structure of the normal protein was described in 1996 and 1997 by Rudolf Glockshuber and Kurt Wüthrich and their respective groups in Zurich, making use of nuclear magnetic resonance techniques. This structure differed in a number of ways from that suggested by Cohen, Prusiner, and their colleagues. It contained three alpha-helices and two small beta-sheets; but most important, the first half of the chain—about a hundred amino acids in length—was unstructured. It was suggested that this long unstructured region could become structured with relative ease and without requiring the unspooling of the remainder of the molecule, to yield the beta-sheet-rich infectious form. To verify this, however, we would need to know the structure of the infectious form, which, as of this writing, we do not.

Still, it had been established that the three-dimensional structure of the infectious protein differed from that of the normal form. We had two molecules containing the same amino acids, linked by the same chemical bonds, but differing in the spatial position of these amino acids—and one was infectious and the other not. It was an eerie echo of Pasteur, the founder of "three-dimensional" chemistry—stereochemistry—and his work on tartrates.

Now that the prion gene had been identified in uninfected cells, two questions arose: What was the usual function of this protein? And, would an animal lacking it still be susceptible to infection?

To get the answers, scientists needed mutant mice or hamsters in which the prion gene had been destroyed. Techniques for producing them had recently been developed, and in 1992 Weissmann's group obtained mice in which the two copies of the prion gene had been inactivated and which were thus completely without the corresponding protein. This showed that the protein was not essential for life. Moreover, no anomalies were detected in the physical or behavioral development of the mutant mice or in their reproductive functions. The role played by the normal protein in uninfected animals was thus unclear, and in fact remains the subject of ongoing research.

In 1993 these mutant mice would provide a striking confirmation of the role of the prion protein. Animals lacking the normal protein did not contract scrapie when injected with the infectious prion. Absent the normal protein, the infectious protein could not be formed to cause the disease.

As with most of a mouse's other genes, two copies of the prion gene were found in each cell, apart from the gametes, or reproductive cells (ova and spermatozoa). But in these mutants, both copies of the gene had been inactivated. The Weissmann and Prusiner groups also bred mutants in which only one copy of the gene had been inactivated or, alternatively, into which additional copies had been introduced. It turned out that the more copies of the gene, the shorter the incubation period. To the extent that the rate of normal-protein synthesis was more or less proportionate to the number of copies of the gene, that result suggested that the infectious protein was formed at a pace closely related to that of the normal protein. That concurred with earlier results showing that the infectious protein was derived from the normal form. The more normal protein there was, the more infectious protein could be produced.

On the basis of the prion gene's presence in the body independent of any infection, it was possible to suggest an explanation for one of the mysteries of scrapie and related human diseases: the absence of an immune reaction to the causative agents of such diseases. The job of the immune system is to recognize foreign substances, whether they be viruses, bacteria, molecules, or other intruders. To do this, it must first be able to recognize what is *not* foreign—the "self"—so that the body will not destroy itself through an autoimmune reaction. As it develops, the immune system learns to recognize the components of the self and becomes tolerant of them. Now, even when it takes on a special structure in its infectious form, the prion protein is basically a component of the self. The body therefore tolerates it and does not initiate an immune reaction. That theory was bolstered by the fact that a significant immune reaction, with substantial antibody production, had been triggered by injecting prion protein into mutant mice that lacked that protein's gene; for these mice, the prion protein was no longer part of the self.

The early 1990s saw the emergence of a theory about the mechanism by which prions multiplied: that the infectious protein, through a kind of "kiss of death," was able to transform the normal protein into its infectious form merely through contact with it. Underlying that somewhat paranormal-sounding theory was John Griffith's second hypothesis, which was in turn based on the Pasteurian school's thinking about the regulation not of protein synthesis but of protein activity.

This had been set out in 1963 in a famous article by Jacques Monod and Jean-Pierre Changeux, both of the Institut Pasteur, and Jeffries Wyman, a researcher working in Rome. It related to the mechanisms involved in enzyme activation and inhibition, phenomena that had a key role in cellular regulation. For example, various cellular compounds could inhibit the activity of the enzyme that was involved in their synthesis, which made it possible for that compound's concentration in the cell to remain constant. In the presence of the inhibitor, the enzyme switched from an active conformation, or shape, to an inactive conformation. Such shape differences appeared to be the result of subtle changes in the way the amino acid chain was folded. The view of Monod, Wyman, and Changeux was that the protein existed in two forms, active and inactive, that were in a state of equilibrium. In a kind of oscillation, the protein constantly switched from one form to the other. By binding to the inactive form, the inhibitor "froze" the protein into that shape. The authors focused in particular on the very common case of enzymes that were composed not of one but of several amino acid chains: dimers, trimers, tetramers, and so forth. The results of their experiments suggested that when the inhibitor bound to one of these amino acid chains, thus freezing it into its inactive conformation, the other chains were automatically frozen in the same form. In a tetramer, for example, when the inhibitor bound to just one of its four amino acid chains, the result was that all four would shift to the inactive form. A single chain would impose its conformation on the other three.

In 1967, considering the possibility that the scrapie agent might be a protein, Griffith proposed the theory that the toxic form of the molecule—a dimer—caused the formation of another toxic dimer by contact with two of the protein's harmless monomers; here he was reviving the ideas of Monod, Changeux, and Wyman. His hypothesis assumed that the conformation of the amino acid chain differed depending on whether it was in its free, nontoxic, state or was part of a dimer, and toxic. In the final analysis, the toxic protein imposed its shape on the normal protein.

That was the theory ultimately embraced in the course of the 1990s by the majority of scientists specializing in scrapie and other spongiform encephalopathies: that the infectious protein converted the normal protein into the infectious form by contact. Such contact could take place in a number of ways. In the simplest variation—which truly deserved the "kiss of death" label—an infectious-protein monomer would come in contact with a normal-protein monomer to form a dimer, within which the infectious protein would impose its conformation on the normal one. The dimer would then split, giving rise to two infectious proteins. Repeated indefinitely, this process would cause an exponential increase in the quantity of infectious protein. A variation would be if the infectious form were a small prion polymer (a tiny fragment of the tubules or fibrils that form both in solution and in the brain). By coming in contact with such a structure, the normal protein would be stabilized in its infectious form. The resulting small polymers, by means of occasional splitting, would give rise to new "seeds" that could stabilize other molecules in their infectious form. This variation of the process was related to crystallization triggered by crystalline seeds in a saturated solution.

This theory, in one form or another, would account for many aspects of transmissible subacute spongiform encephalopathies (TSSEs), some of them already mentioned and others that will be discussed in due course. Here, one point deserves immediate attention: the theory that each prion is specific to one species, the so-called species barrier that has so often been evoked during the ongoing mad cow crisis.

That specificity exists, but it is relative. To be sure, the species barrier sometimes seems to be impenetrable. But in most cases transmission between mammalian species is possible, if difficult. We have seen, in sheep-to-goat, goat-to-mouse, and human-to-chimpanzee transmission, that incubation periods were very long and doses needed to bring about infection were high. But once the agent had adapted to a new species, the incubation period became shorter, and the agent's properties could change. The Prusiner and Weissmann groups studied this in cases of transmission between mice and hamsters. It was difficult to transmit scrapie between those two species, and incubation periods were long. The two research groups constructed inbred lines of mice whose prion gene had been replaced with that of a hamster. These mice behaved like hamsters vis-à-vis prion infection; they proved to be highly sensitive to the hamster prion and highly resistant to the mouse prion. The specificity thus lay in the structure of the prion itself.

This was confirmed by the construction of additional, more complex, inbred lines of mice, such as lines possessing both mouse and hamster prion genes. When these were infected with the hamster prion, they produced hamster prion, which could easily infect hamsters but could infect mice only with difficulty. And when the mice were infected with the mouse prion, they produced mouse prion. These studies showed that the prion used for infection "preferred" to contact the normal protein of the same species rather than that of a different species. The "kiss of death" took place more naturally between proteins deriving from a single species than between those deriving from different species.

This fact could be melded into the theory mentioned earlier to account for prion multiplication. It assumes contact between infectious and normal proteins. Such contact would be relatively easy between proteins sharing the same amino acid sequence, but it could be more difficult if the sequences were not absolutely identical, as in the case of proteins from different species. Interaction would take place with difficulty in the case of interspecies transmission, which would account for the lengthy incubation periods.

An explanation was still needed for the adaptation that occurred once the infection took hold. One possibility was that during the initial interaction with a prion from another species, the infectious conformation adopted by the infected animal's protein would be that which was specific to that animal. Beyond its theoretical interest, this matter would have great practical importance when the causes and consequences of the mad cow crisis came to be examined.

But we are not yet at that point in our story. We must first see how the new theories bore on an issue that has constantly haunted specialists both in scrapie and in human diseases of the same kind: Are these diseases infectious or genetic? It might seem that the question had been settled. These were infectious diseases, and the causative agents, while very unusual, had been identified. They were the proteins known as prions. And yet . . .

18

THE RETURN OF THE SPONTANEISTS

SCRAPIE IS TRANSMISSIBLE. We have known this since the 1936 experiments of Cuillé and Chelle and the accidental contamination of thousands of sheep in Scotland by a louping ill vaccine around the same time. Human spongiform encephalopathies—kuru and Creutzfeldt-Jakob disease—are infectious as well, as Gajdusek's team demonstrated in the late 1960s by transmitting them to chimpanzees. So if you accept the orthodox Pasteurian "contagionist" view, these diseases can be contracted only through contagion.

If spongiform encephalopathies were infectious, the next task was to find out how they were transmitted. For scrapie, there were very convincing arguments in favor of natural contagion, and at least one possible method of transmission had been suggested: through the consumption of infected placentas. For the human diseases, hardly any doubt remained about the transmission of kuru through cannibalism. But what about Creutzfeldt-Jakob disease, which, unlike kuru, affected the entire human population? Occasional examples of human-to-human transmission had been identified before 1985; these were cases of iatrogenic transmission, such as those described in Chapter 14. Additional cases would subsequently result from human growth hormone treatment and from dura mater transplants.[1] But these accounted for only a very small percentage of CJD cases worldwide. What had caused all the other cases? Were they also a result of contagion?

Another theory began to emerge toward the end of the 1980s following study of Gerstmann-Sträussler syndrome (GSS)—also known as Gerstmann-Sträussler-Scheinker syndrome, although Scheinker's name is commonly omitted—a disease that had barely been mentioned since its discovery in 1936. The connection between GSS and CJD was made only belatedly, in the early 1980s. Although the diseases were similar, there were major differences. Gerstmann-Sträussler syndrome affected younger patients—the average age at death was forty-eight, compared with fifty-eight for Creutzfeldt-Jakob disease—and once the disease was manifested it lasted far longer—about five years rather than six months. There were other differences as well, relating to the clinical symptoms and to the nervous system lesions. For example, the brains of GSS patients generally showed many amyloid plaques with properties somewhere between those seen in kuru and those seen in Alzheimer's disease, but these plaques were rarely seen in patients with CJD. However, exhaustive study of several cases of Gerstmann-Sträussler syndrome and the wide variety of guises in which Creutzfeldt-Jakob disease could appear led to the conclusion that GSS was actually a somewhat unusual form of CJD: The amyloid plaques seen in GSS patients contained prion protein; and, like CJD, this syndrome could under some circumstances be transmitted to primates.

So Gerstmann-Sträussler syndrome proved to be The Disease in yet another of its disguises.

Gerstmann-Sträussler syndrome was a very rare disease, affecting barely more than one person in a hundred million each year. Even when sporadic cases were observed, they often had a hereditary element. In affected families, the disease struck an average of half the individuals in a given generation. Here, one could still blame congenital infection, like that previously seen in tuberculosis. But a number of arguments favored the theory that this was a genetic disease caused by a dominant mutation, as Gerstmann, Sträussler, and Scheinker had originally postulated. If GSS was indeed a genetic disease, and in light of the key role played by prions in such diseases, perhaps GSS resulted from a mutation in the prion gene. That was indeed the conclusion that Prusiner and his colleagues reached in 1989. In one American and one British family, both affected by Gerstmann-Sträussler syndrome, they had identified the same changes in prion-gene base pair sequence, leading to a change in the nature of the protein's 102nd amino acid. This change was specific to GSS. It was found neither in a sample of one hundred people drawn from the population at large nor in fifteen patients suffering from other forms of Creutzfeldt-Jakob disease.

Gerstmann-Sträussler syndrome thus seemed to be a genetic disease caused by a mutation in the prion gene. But it could be transmitted to animals and was thus also infectious.

That finding would unleash a flood of work showing a prion gene mutation in the known cases of Creutzfeldt-Jakob disease that either were genetic or displayed an abnormally high prevalence in a given geographic region. This mutation was the same within a family or within a region, but it generally differed from family to family or from region to region. Scientists thus found a score of mutations associated with various forms of Creutzfeldt-Jakob disease. I shall cite two of them.

The first was a new form of The Disease first described in 1986; like the others, it was initially considered to be a new disease and was named "fatal familial insomnia" (FFI). As the name suggests, its first symptom was loss of sleep—initially partial, then total—accompanied by increasingly frequent episodes of hallucinations. Death came in less than a year. Neither the clinical symptoms nor the nervous system lesions suggested Creutzfeldt-Jakob disease. But once again, it was indeed a variant of CJD. As with Gerstmann-Sträussler syndrome, a protease-resistant form of prion protein was found in the brains of patients, and the disease could be transmitted to animals. Patients had a prion gene mutation, with an amino acid change at position 178. That mutation was specific to fatal familial insomnia: It was found in all tested patients displaying its symptoms. But the opposite was not true, because people with that mutation could also display symptoms closer to those of classic Creutzfeldt-Jakob disease. Here, as in other instances, the conclusion was that the nature of the mutation did not completely determine the characteristics of the disease.

The second example was one we have already mentioned: that of Jews of Libyan descent living in Israel. CJD patients in that group had been found to carry a prion gene mutation, in this case involving an amino acid change at position 200. The same mutation was later found in other Jewish families of Mediterranean origin living outside Israel. Genealogical study suggested that the gene carrying the mutation came from a distant common ancestor who might have lived on the Tunisian island of Jerba. This discovery laid rest to the notion that the high prevalence of CJD in this group was a result of diet. The same mutation was then discovered in several other countries, including in central Europe and South America. This made it possible to trace it back to Sephardic Jews who had fled Spain and the Inquisition in 1492.

In the cited cases of genetic disease, the mutation was generally present in only one of the two copies of the prion gene. Only that single mutated copy was sufficient to cause the disease, which indicated that the mutation was indeed dominant, as suggested by genealogical studies. On the other hand, the mere presence of the mutation was not always enough to bring about the disease. A mutation at position 102 meant that an individual would almost certainly fall victim to Gerstmann-Sträussler syndrome (barring premature death from another cause), while Libyan Jews with an amino acid change at position 200 were likely to live to an advanced age without contracting Creutzfeldt-Jakob disease.

So, Creutzfeldt-Jakob disease in its various forms was simultaneously infectious and hereditary. What could this paradox mean?

Let us go back to the prion theory, according to which the infectious protein is an abnormal form of a protein that is ordinarily present in the brain. The normal protein shifts to the abnormal conformation upon contact with the infectious protein. Under that theory it is easy to conceive that some protein gene mutations would cause the protein spontaneously to assume the pathogenic, infectious form; the probability of this would vary according to the mutation. This would explain how in some instances nearly all people with the mutation would be affected by the disease, while with other mutations there was a chance that no illness would appear.

The interpretation that had been suggested with respect to inherited Creutzfeldt-Jakob disease could also be adapted to fit the far more frequent cases of sporadic Creutzfeldt-Jakob. The first thing that comes to mind is that these cases too are connected with prion gene mutations. If these mutations did not appear until the ovum or spermatozoon was formed in the patient's parents, or even until the first stages of embryonic formation, the patient's forebears would not have been carriers, and the disease would thus not be hereditary in nature. But analysis of the prion gene in many sporadic CJD patients eliminated that theory. Those patients showed no prion gene mutation.

Another, similar, idea was that a mutation came about during an individual's development, so that only daughter cells of cells with the mutation would carry that mutation. If such a mutation—known as a somatic mutation—occurred during embryonic development in a cell that was to give rise to the nervous system, the nervous system alone would have the mutation, not the body's other tissues. So, even though most of an individual's cells were without the mutation, it would be enough for a small number of mutated cells to synthesize a small quantity of infectious protein, because this would, step by step, convert the normal protein synthesized by the other cells into the infectious form.

Yet another theory suggests that there was a possibility—a very small one, but not nonexistent—that a normal protein could shift to the infectious conformation. When this happened (in about one person in a million), it would trigger a chain reaction: The wrongly formed protein would progressively contaminate a substantial number of normal molecules, as occurs during an infection. In this scenario, mutations present in the inherited forms of the disease would simply increase the probability that the protein would spontaneously adopt an infectious conformation. This theory could explain how a disease could be simultaneously infectious, spontaneous, and hereditary—vindication for the spontaneists who had argued against Pasteur in the nineteenth century.

Was it vindication also for the farmers and veterinarians who had viewed scrapie as a hereditary disease? There, the situation was not quite so clear. Let us briefly review it.

When scrapie was first described in the eighteenth century, many farmers thought it a hereditary disease. That idea was abandoned in the wake of subsequent proof that it was caused by an infectious agent, along with observations showing that it was contagious. Cases claimed to be hereditary or spontaneous were seen as the result of undetected contamination. Now that it appears Creutzfeldt-Jakob disease can be either inherited or spontaneous, we may well wonder whether the same might be true for scrapie. To put it another way, even though it seems likely that a large percentage of scrapie cases within flocks are the result of contagion, could not other cases be spontaneous or inherited?

That question has yet to be answered satisfactorily. All we can say is that some countries, such as Australia, seem to have long been scrapie-free. That would not be so if there had been cases of sporadic scrapie, as there are of sporadic Creutzfeldt-Jakob disease. But this argument remains comparatively weak, because isolated cases of scrapie could go undetected in the absence of contagion to other members of the flock.

So it is not known whether there is any such thing as sporadic or inherited scrapie among animals, as there is for CJD in humans. On the other hand, we know for a certainty that sensitivity to scrapie infection is determined by genetics, because exhaustive observations have been made among mice.

When he carried out the earliest goat-to-mouse transmissions (see Chapter 7), Richard Chandler observed that not all mouse inbred lines were equally receptive. A few years later, this was the focus of another group of British researchers, which included Alan Dickinson. In 1968 that group identified a gene controlling the disease's incubation period in mice. They observed that in one of the mouse inbred lines they were working with, the incubation period was thirty-seven weeks, while in five other lines it ranged from twenty-one to twenty-six weeks. When they crossed the long-incubation inbred line with one of the shorter-incubation lines, they observed that this behaved like a genetic trait: A trait for the incubation period of scrapie was carried by a gene. Genetics, then genetic engineering, would ultimately establish, in 1998, that this was none other than the prion gene. There were small differences in sequence between the prion genes of different mouse lines, and these affected the animals' sensitivity to inoculation with scrapie. Similar results were obtained among sheep, where considerable polymorphism relating to the nature of several of the protein's amino acids gave rise to a whole gamut of sensitivity to scrapie. That explained the varying levels of sensitivity seen in different breeds of sheep.

Dickinson's experiments also uncovered another fact relating not to mouse inbred lines but to prion strains. We saw in Chapter 7 that Pattison and Millson had found two strains of the scrapie agent in goats, causing "drowsy" and "scratching" symptoms respectively. Dickinson in turn identified several strains that differed in such areas as the incubation period of the disease they caused. He used two such strains to inoculate two lines of mice that themselves displayed different incubation periods. The outcome was not exactly what he had expected: A line of mice that displayed a short incubation period with one prion strain displayed a long incubation period with the other, and vice versa. The protein sequence differences between the two lines of mice did not correspond to absolute differences in sensitivity to infection, but rather to sensitivity differences that were specific to a given prion strain. We shall return to the question of prion strains in Chapter 24; as it is difficult to integrate into prion theory as we have been describing it thus far.

But now, let us sum up the prion theory, which is today accepted by the majority of specialists in the field.

Transmissible subacute spongiform encephalopathies (TSSEs)— which include scrapie in animals and Creutzfeldt-Jakob disease in humans—are caused by a protein known as a prion, a nonpathogenic, "normal" form of which is found in all mammals. Prions can adopt an abnormal pathogenic and infectious form that apparently relates to a difference in the folding of the amino acid chain. The shift from normal to infectious form can very occasionally occur spontaneously; this is sporadic Creutzfeldt-Jakob disease. The presence of certain mutations in the prion protein gene increases the probability of such a shift; these are the inherited forms of CJD. The infectious form of the protein is able to compel the normal form to adopt the infectious conformation; this is what happens during natural contagion of scrapie within a flock, during iatrogenic contamination with CJD, and during the transmission of kuru through cannibalistic practices. That capacity of the infectious protein to turn the normal protein into its infectious form also explains the way in which The Disease spreads through the body once the shift from one form to the other occurs in a cell, either spontaneously or as the result of infection.

As you can imagine, this theory turned many widely held ideas on their heads. It did not explain everything, but it seemed to hold at least a substantial part of the truth—as the Nobel committee recognized in 1997 when for the second time it awarded a prize for work on TSSEs. The laureate was, of course, Stanley Prusiner, who had introduced the concept of the prion.

That concept had been launched, and named, in 1982, but it was not really fleshed out until 1985, when it was demonstrated that the prion gene was present in all mammals. Prusiner's Nobel prize twelve years later reflected the fact that the concept had gained enough strength to be accepted by the majority of the scientific community, despite its initial tinge of heresy.

By the time Prusiner received the prize in 1997, we might have hoped to be celebrating this new victory of the human mind. But unfortunately, the fact was that, while scientists were taking giant steps forward in their understanding of The Disease, The Disease itself was taking advantage of medical and veterinary progress to claim new victims. The fears that arose in April 1985 had become reality: The number of patients who had contracted CJD through contaminated human growth hormone was continuing to increase, and the "mad cow" epidemic had developed into a catastrophe for farmers and a matter of grave concern for human health.

19

TO GROW—AND TO DIE

GROW . . . AND DIE: That was the sad fate of at least 140 children treated with human growth hormone and infected with the causative agent of Creutzfeldt-Jakob disease. The final outcome of this tragedy remains to be known; owing to the lengthy incubation period of CJD, additional cases appear every year.

The announcement of the first cases, in April 1985, took specialists by surprise; the vast majority of them had not seriously contemplated the possibility of such infection. Although in the period 1974–1977 Gajdusek and his team had signaled the risk of transmitting the CJD agent during neurosurgery, organ transplants, and even blood donations, no scientific publication prior to April 1985 had mentioned such a risk in connection with human growth hormone. It later turned out that there had been some concern within the British Medical Research Council; later in this chapter we will describe how they tried to assess the risk. The question had been raised in France as well, in 1980, in circumstances that are worth recounting.

Toward the end of 1979 a cornea transplant in a French hospital gave rise to an iatrogenic infection from the rabies virus. As this was the first human case of rabies in France for many, many years, it gave rise to a great deal of talk. It also aroused concern within L'Association France Hypophyse—the French Pituitary Association—that viral infections could also come about through human growth hormone treatment, and the association sought the advice of a virologist. It turned to Luc Montagnier of the Institut Pasteur (who was yet to discover the AIDS virus). In his reply,[1] Montagnier made reference to Creutzfeldt-Jakob disease, with which few members of the association were likely to be acquainted:

> The threat of contamination of the hormone preparation can thus come from subjects who died of acute neurotropic viral disease . . . or of slow virus encephalopathies, among which are . . . unconventional agents that are still poorly defined (such as Kreutzfeldt-Jacob [*sic*] disease)[2] or that are simply putative (such as multiple sclerosis and Parkinson's disease). . . . Special attention should be paid to the danger of transmitting Kreutzfeldt-Jacob disease (KJ), a rare disease to be sure—an average of one case per million—but of which carriers may be far more numerous. The infectious agent, which is similar if not identical to those of kuru and scrapie, is highly resistant to heat, denaturing agents, ionizing and non-ionizing radiation. . . .
> I suggest that, for the moment, preventive measures be taken with a view to lowering the above-mentioned risks: to eliminate as pituitary donors:
>
> · All subjects who died of an acute neurotropic-viral disease. . . .
> · All subjects who died of a viral or non-viral encephalopathy. . . .
> · All subjects who had exhibited serious, rapidly developing neuropsychiatric problems in the two years preceding their death (which could come about from a different cause). This would make it possible to eliminate patients who were Kreutzfeldt-Jacob carriers but who died of another cause before the disease had matured (which rarely takes more than eighteen months), but not of slower-developing diseases such as multiple sclerosis or Parkinson's disease—whose viral origin, in fact, has not been proven.

In the future, a study could be considered comparing the effects of very high doses of gamma radiation on the KJ agent and on the

growth hormone itself. It is to be hoped—though it is not certain—
that selective inactivation of the infectious agent would be achieved.

The administrative council of L'Association France Hypophyse met
a few days later to decide what action to take on Montagnier's report.
Following that discussion, the founding president of the association, the
great pediatrician Pierre Royer, recalled that "to date, among nearly six
hundred children treated [with human growth hormone] in France and
systematically monitored quarterly, no accident or incident has been
noted. The risk, if it exists, is thus very minor. Nonetheless, the goal
should be total safety." He therefore suggested following Montagnier's
recommendations on the harvest of pituitaries from cadavers, and this
was endorsed by the administrative council. The council decided to
issue a directive to hospitals carrying out such harvesting about the
measures they were to take. Moreover, it called for experiments on
whether ionizing radiation—gamma rays—could inactivate the CJD
agent without destroying the growth hormone. Such experiments were
carried out after consultation with the prominent radiobiologist Raymond Latarjet, who had earlier studied the radiation sensitivity of the
scrapie agent. The outcome was inconclusive. When pituitaries were
subjected to the quantity of radiation that would have been necessary,
they no longer yielded any active growth hormone.

New regulations on pituitary harvesting were formulated on the
basis of the administrative council's recommendations, and the preparation and distribution of human growth hormone resumed as before.
The risk of CJD transmission would not be mentioned again until April
1985. A 1983 inquiry by the Inspection Générale des Affaires Sociales
objected to certain administrative and financial practices related to pituitary harvesting, but made no mention of the possible risk of transmitting infectious diseases. The words "Creutzfeldt-Jakob disease" did not
appear anywhere in that report. Not only L'Association France Hypophyse but all the other organizations and firms producing human
growth hormone worldwide were confident in the quality of their

product, which was obtained through a multistage purification process. And, contrary to what many believe today, that confidence did not evaporate overnight when the first probable cases of CJD contamination were reported in April 1985. In fact, an editorial in *The Lancet* announcing the ban on the use of human growth hormone in the United Kingdom stated:

> The extraction procedure of the US preparation of HGH differs
> from the one in the UK and the risk attached to the latest UK
> preparations is thought to be slight. As with any extraction proce-
> dure, great improvements have been made over the years and the
> hormones prepared in this way are very much more pure than they
> were initially. Nevertheless, since it is impossible to guarantee ab-
> solute safety, it was thought best to place a ban on HGH supplies
> and wait for biosynthetic HGH to become available, probably
> within the year.[3]

That was then the belief of many doctors not only in the United Kingdom but also in France and elsewhere. They felt that there might have been some contamination—although some doubted even this—but that it had occurred in the late 1960s or the very beginning of the 1970s, when the hormone in use was not very pure. It seemed highly improbable that such an accident could recur, given the considerable improvement in purification methods. Indeed, the British had a particular reason to think this.

As noted above, some Medical Research Council scientists had been concerned about this in the late 1970s. To get to the bottom of the problem, they decided to simulate the contamination of their pituitaries and find out whether their purification method was able to eliminate the infectious agent. Rather than using the CJD agent, which was difficult to measure and dangerous to handle, they used the scrapie agent. They added a preparation of this to pituitary extract, then proceeded to purify growth hormone from that extract and to measure the quantity of agent in the purified fractions by inoculating mice. The results were known in

1983 but not published until August 1985: The researchers could detect no infectious capability in the purified hormone. Their purification method, which was used also in France and elsewhere, thus seemed able to strip the hormone of any possible contamination by the CJD agent. Still, the authors stressed the need for the greatest attention to detail in all steps of extraction and purification. For example, it was important to avoid recontaminating a purified fraction with glassware that had not been subjected to the special sterilization process needed to eliminate this kind of infectious agent. But, although the risk of transmission seemed very small, it was not nil, and measures were needed to reduce it further or to eliminate it.

As we have seen, the United States and the United Kingdom, followed by some other countries, decided immediately to ban the use of growth hormone extracted from human pituitaries. Despite the risks, it would surely have been more difficult to reach that decision if such a ban had meant that a treatment highly valued by patients and their families would never again be available. But luckily, in fact, a new method of hormone production was about to emerge. It was the result of work carried out in the late 1970s at the Institut Pasteur in France and at the University of California in the United States. Those experiments, whose initial outcome was published in 1979, involved introducing the human growth hormone gene into a bacterium in order to "force" it to synthesize the hormone. Beyond its scientific interest—these were the early days of genetic engineering—this work had the primary goal of finding an unlimited source of human growth hormone; safety was not a consideration. Quite the contrary, in fact: In 1979 most people viewed genetic engineering with suspicion; there was much concern about the use of products obtained through such technology. But by 1985 views had changed, and it seemed that human growth hormone produced by a bacterium would solve the problems that had arisen with hormone extracted from cadaveric pituitaries. It turned out that one of the first of the American biotechnology companies, Genentech, had begun work on this hormone in 1979 and was on the verge of marketing it. This coincidence—which proved to be very fortunate—was rather puzzling at the time. Some people believed that the publicity surrounding the few—and sometimes dubious—cases of Creutzfeldt-Jakob disease in children who had been treated with cadaveric growth hormone was intended to prepare the ground for the introduction of so-called biosynthetic or recombinant human growth hormone. Be that as it may, in 1986 the sale of recombinant hormone began in the United States; by 1988 this had completely replaced hormone derived from pituitaries nearly everywhere in the world.

While the United States, the United Kingdom, and a few other countries had stopped using cadaveric human growth hormone by the spring of 1985, that was not the case everywhere. Specifically, France had decided to continue to use such hormone, but to alter the purification technique with a view to eliminating or minimizing the risk of contamination. The key modification was to treat the purified hormone with concentrated urea. Urea was known to unspool proteins—to unwind the "ball" formed by the amino acid chain during synthesis. Once unwound, this ball would re-form only with difficulty—or not at all—when the urea was removed. During research on the nature of the scrapie agent, it had been noted that concentrated urea would destroy its infectious power. Moreover—as those responsible for human growth hormone preparation in France would not learn until April 1985— some pharmaceutical companies used urea in preparing the hormone. It was used not to protect against the possibility of contamination but to dissociate the aggregates that the hormone sometimes formed, and thus to increase yield. But this was a most important fact, because it showed that, unlike most proteins, this hormone was not inactivated by urea, which could thus be used to do what gamma-ray irradiation had been unable to do: selectively inactivate the CJD agent, which was very similar to that of scrapie, without inactivating the hormone.

From May 1985 until its replacement with biosynthetic hormone in 1988, human growth hormone produced in France and in several other countries was thus treated with urea. There were a number of reasons behind France's decision to continue distributing hormone extracted from pituitaries. Among them was uncertainty about when biosynthetic hormone would be available, along with a certain mistrust of that product. This mistrust arose in particular from the fact that the first biosynthetic hormone on the market was not absolutely identical to natural hormone. It contained an additional amino acid at one end. There was therefore some fear that the patient's immune system would identify it as a foreign body and reject it. This had occurred in some diabetics treated with insulin derived from cows or swine, which was slightly different from human insulin. In any event—and however we may judge them in retrospect—the methods used in different countries were equally effective: No case of Creutzfeldt-Jakob disease known as of 2001 can be attributed to human growth hormone treatment initiated after June 1985. Unfortunately, for children treated before then, the damage had already been done.

Yet for three or four years it seemed that the four cases reported in April 1985 were anomalous accidents. Perhaps they could be attributed to the contamination of a single lot of pituitaries whose hormone had been purified using old, now-abandoned methods. True, three further cases were reported in 1987 and 1988, but that made only seven cases out of about twenty-five thousand children treated with human growth hormone. This was hardly an epidemic. And the facts seemed to vindicate those who had stated their confidence in the hormone preparation process. But, sadly, the situation would change in 1989 and 1990. By 1990, the total number of deaths from CJD attributable to human growth hormone had risen to thirteen: seven in the United States, four in the United Kingdom, one in New Zealand, and one in Brazil. Would France be spared? Unfortunately not. Quite the contrary.

In 1992 France reported four cases of CJD: in two children aged ten and eleven, and in two young adults aged eighteen and nineteen. The first symptoms had appeared six to twelve years after beginning the human growth hormone treatment. The number of cases would continue to rise, especially in France. By the end of 2000, the total was 139, of which seventy-four were in France. Obviously, it was no longer a question of one tainted lot of pituitaries, or of a particular method of preparation. Several lots had to have been contaminated, and modern purification methods—until the introduction of urea treatment, which had solved the problem—had been no more effective than previous ones in completely decontaminating the hormone. The level of contamination seems to have been higher in France than elsewhere, as reflected in the greater number of victims and in a shorter incubation period. A retrospective epidemiological analysis shows that the young French patients had been infected during a critical period between late 1983 and May 1985. There thus had to have been a major contamination during that period, the reasons for which are not known with certainty.

For the children and young adults who died, and for their families, this was a tragedy. For those who had been treated with human growth hormone—especially the thousand or so young French patients who received it during that critical period and who still remain in good health—it is a source of dread.

20

LESSONS LEARNED

COULD THE TRAGEDY have been averted? In theory, perhaps, but only at the cost of abandoning the treatment of pituitary dwarfism.

By the late 1960s—even before the foundation of L'Association France Hypophyse—the scientific literature contained information that could have signaled the danger of administering growth hormone derived from human pituitaries. But pediatricians and endocrinologists treating pituitary dwarfism either did not know of that information or did not fully appreciate it. Remember that it was in 1968 that Gajdusek's team succeeded in transmitting Creutzfeldt-Jakob disease to chimpanzees, thus demonstrating its close resemblance to scrapie and gathering both diseases together under the label "subacute spongiform encephalopathies." It seemed that, although of unknown nature, the causative agents of these diseases had to be very similar—so what was true for scrapie had a good chance of being true for Creutzfeldt-Jakob disease as well.

And veterinary researchers had obtained findings that could have given pause to those in charge of treating pituitary dwarfism. In 1963, Chandler showed that the brain of a single scrapie-infected mouse could infect millions or even billions of other mice. In the early 1950s, Pattison and Millson had sought the scrapie agent in various organs of infected goats, and had found it in the pituitary. And the fact that scrapie could be transmitted by intramuscular means had been known since 1936, when several hundred sheep were accidentally infected through a vaccine against louping ill (see Chapter 6). By the end of the 1960s there was thus every reason to suspect that Creutzfeldt-Jakob disease could be transmitted by means of hormones extracted from pituitaries. But what pediatrician or endocrinologist was likely to be immersed in the veterinary literature, some of it dating back thirty years or more? And one key piece of information was lacking for assessing the risk: the frequency of the disease in humans.

The first epidemiological report on frequency was published only in 1979. This was the study that gave rise to the famous figure of one case per million per year, which was subsequently confirmed, and was cited by Montagnier in his January 1980 letter to L'Association France Hypophyse (see Chapter 19). That eye-catching, easy-to-remember figure marked Creutzfeldt-Jakob as an extremely rare disease, which made the chances of harvesting a contaminated pituitary appear very slim indeed, even if, like Montagnier, you took into consideration any possible healthy carriers. But this was to forget that pituitaries were not harvested from the population at large, but only from people who had died, among whom cases of CJD—which is quickly fatal—were far more frequent than among the overall population. Brown, Gajdusek, and their colleagues noted this point in an article published in September 1985, following the first cases of contamination:

> The U.S. annual mortality rate from all causes during the 1960–1980 period was approximately 0.9 percent, or, in the population of 250 million, somewhat fewer than 2.5 million deaths each year. Since the annual mortality rate from Creutzfeldt-Jakob disease is approximately 0.7 to 1.0 per million, or, in the U.S. population, somewhat fewer than 250 deaths per year, it follows that roughly 1 in 10,000 deaths in this country is due to this disease. Because lots of pituitaries used in the preparation of human growth hormone have varied

from 500 to nearly 20,000 glands, frequent episodes of contamination can be expected to have occurred, unless patients with
Creutzfeldt-Jakob disease were systematically excluded as
sources of pituitary glands.[1]

The authors went on to say that such exclusion had not been and
could not be carried out in a truly effective way. To the contrary, the
proportion of pituitaries from persons affected by the disease probably
exceeded one in ten thousand, because upon death such patients were
more likely to be subjected to autopsies including trepanation, which
made it easier to remove the gland. In other words, it was very likely that
a considerable proportion of the lots of pituitaries used worldwide to extract human growth hormone included at least one contaminated gland.

The special properties of the CJD agent—its tendency to "stick" to
everything and to form aggregates of varying size, its extraordinary resistance to the usual decontamination methods, and the impossibility of
measuring it—made eliminating it during the hormone purification
process a very chancy business. Brown, Gajdusek, and their colleagues
were thus right to envision from the outset the possibility of a massive
epidemic among young people who had been given human growth hormone therapy. But in fact, although there have been, unfortunately, a
large number of cases of contamination—about 140 worldwide as of
2001—it seems likely that the epidemic will remain limited and that
only a very small fraction of young people treated with human growth
hormone will have been affected. There are a number of reasons why
this epidemic has not been larger: Purification methods succeeded in
eliminating a large proportion of the infectious agent; intramuscular infection is not very effective; possible variations in individuals' sensitivity
to the disease; and, perhaps, the infectious agent's tendency to form aggregates, which could concentrate it in clumps in only a few of the vials
of hormone yielded by each purification procedure.

The unpredictable nature of contamination from lot to lot was illus-
trated by American attempts to identify the contaminated lot or lots of

hormone. In 1985 researchers injected monkeys with samples from seventy-six different lots; each lot was used to inoculate three monkeys. After five and a half years, only one monkey of the 228 inoculated had come down with Creutzfeldt-Jakob disease; even the two other monkeys inoculated from that same lot remained unaffected.

Let us imagine for a moment that, following Montagnier's letter, hormone producers had wished to completely eliminate all risk. Their only alternative would have been to immediately halt all treatment for pituitary dwarfism, because no method of donor selection would have enabled them, for a certainty, to prevent the inclusion of an infected pituitary in the lots from which the hormone was extracted. Moreover, in the absence of a biochemical or immunological method of measuring the CJD agent, there was no way to guarantee the safety of the purified hormone. Patients and their families would have found it hard to understand why producers had decided to halt a treatment that had been used for twenty years with complete satisfaction because of a purely theoretical risk from an agent that had yet to be identified.

However distressing it may have been, this misfortune did add to our knowledge of The Disease, with respect to its symptoms and to the genetic predisposition to it. This knowledge was largely the result of work by John Collinge's team in the United Kingdom, and by French specialists in transmissible spongiform encephalopathies: Dominique Dormont and his colleague Jean-Philippe Déslys, and the team led by Thierry Billette de Villemeur.

The clinical symptoms of iatrogenic CJD proved to differ from those of the spontaneous form, known as sporadic CJD. In the iatrogenic disease, the earliest symptoms, such as problems with balance, systematically indicated that the cerebellum had been affected, whereas the first symptoms of sporadic CJD most often involved dementia, which indicated that the brain per se was affected. These clinical differences were correlated to a difference in the location of the infectious protein as detected by its protease resistance. In patients with iatrogenic CJD, large quantities of the protein were found in the cerebellum and smaller quantities in the frontal lobe of the brain. Most commonly, the reverse was seen in cases of sporadic CJD. Moreover, microscopic examination of the brain frequently showed the presence of amyloid plaques in iatrogenic cases, while these were rarely seen in sporadic cases. Broadly speaking, the iatrogenic form recalled kuru. It must be noted that both iatrogenic CJD and kuru involve peripheral infection, either via intramuscular injection or orally, while sporadic CJD originates within the nervous system itself. Some scientists feel that peripheral infection could result in the selection of a particular prion strain differing from the one that causes the sporadic form of the disease.

The other observation brings us back to the prion gene. Certain mutations of that gene are known to be responsible for inherited Creutzfeldt-Jakob disease. Apart from such mutations, the gene is nearly identical in all members of the population—*nearly* identical, but not completely. Like many other genes, this one is prone to a certain degree of polymorphism. That means there can be very slight differences in sequence within the population, reflected in one or more differences in the amino acids contained in the corresponding protein. In the case of the prion protein, the amino acid at position 129 is sometimes a valine (V) and sometimes a methionine (M), which are two of the twenty amino acids found in proteins. Everyone possesses two copies of the prion gene, and these can encode either one protein with the same amino acid at position 129 (such individuals are called homozygotes) or two proteins with different amino acids at that position (heterozygotes). Among the general European population, 51 percent are M/V heterozygotes, 37 percent are M/M homozygotes, and 12 percent are V/V homozygotes. Very different proportions are found in iatrogenic CJD patients. In the French study, an early result, published in 1994, was that all twenty-three such patients whose prion gene had been analyzed were homozygotes. This seemed to show that heterozygotes—who would have accounted for about half of the children treated with human growth hormone—would have been resistant to injection of the CJD agent. Five years later in 1999, the proportion of homozygotes, while not 100 percent, was still 83 percent: forty-six of the fifty-five patients whose prion gene had been analyzed. In the remaining nine (heterozygous) patients, the disease had appeared starting in 1994. Thus, they may not have been completely resistant to inoculation with the CJD agent, but the incubation period appeared to be longer for them than for homozygotes.

Collinge's team also had found a far larger number of homozygotes than heterozygotes among British patients who had contracted CJD as a result of human growth hormone therapy. The team then turned to cases of sporadic CJD and, to their surprise, found the same imbalance. Homozygotes accounted for more than 80 percent of those patients too, far higher than the 49 percent in the population at large. These findings indicated that heterozygotes, whose cells contain two different kinds of prion protein—one with a valine and the other with a methionine at position 129—are generally more resistant to Creutzfeldt-Jakob disease. That outcome recalls the theories offered to explain the species barrier. Perhaps the interaction between prion proteins, thought to be necessary for triggering the process of infection, could take place effectively only between identical proteins. That process could be slowed or even prevented if the cells contained a mixture of two slightly different kinds of protein.

The tragic infection of children treated with human growth hormone was revealed in April 1985, with the announcement of the first cases in the United States. By a strange coincidence, the first cases of what would be identified as "mad cow disease" surfaced at the same time. Initially, there was no link between the two events. But both would yield the same result: the infection of human beings with the causative agent of Creutzfeldt-Jakob disease.

21

HAVE THE COWS GONE MAD?

ALTHOUGH THE FIRST CASES of bovine spongiform encephalopathy were recorded in April 1985, they were not identified as such until two years later. That may seem surprising, but we must recall that a single cow in a herd dying, even of a murky cause, is not so extraordinary as to justify an in-depth investigation. The animal is written off as a loss on the balance sheet, its carcass is sent to the processing plant, and nothing more is said of it. Things become more serious when there are several deaths within a herd from what appears to be a single cause. There is the danger of an epidemic, and it is the veterinarian's duty to inform the health authorities. That is exactly what happened in one southern English county, where between April 1985 and February 1986 nine cows in one herd were stricken with serious nervous system disorders that ultimately led to death or slaughter. In November 1986, the brains of two of these animals were examined at the Central Veterinary Laboratory in Weybridge, Surrey, not far from London. The diagnosis was a spongiform encephalopathy. Between December 1986 and May 1987, lesions of the same type were observed in the brains of four more cows, three of them from three other herds from different counties. There was no longer any room for doubt: Something strange was happening. British veterinarians were informed of this seemingly new disease and were urged to make a declaration when they detected the symptoms in any animal. But the scientific community was not informed until late October 1987, with the publication of an article by Gerald Wells and colleagues from the Central Veterinary Laboratory.

That article was the first to describe, albeit briefly, the symptoms and associated brain lesions of the disease that the authors named "bovine spongiform encephalopathy," or BSE. The lesions were found mainly in the gray matter of the brain stem, where the spinal cord joins the brain, and were characterized by numerous vacuoles that gave the gray matter a bubble-like, spongy appearance—hence, the term *spongiform*. The authors stressed the close similarity with scrapie lesions, especially since in one case they had observed fibrils similar to those found in brain extracts from animals with scrapie. There was thus a strong suspicion that this was a new incarnation of The Disease—a new transmissible spongiform encephalopathy—but this remained to be proved.

The active involvement of the veterinary community in the spring of 1987 would quickly bear fruit; several dozen new cases were reported. Their number and distribution demanded a complete epidemiological study to assess how widespread the disease was and to try to find its causes. Such an epidemiological investigation got under way in June 1987.

The next year marked a turning point. The transmissibility of BSE was demonstrated, its probable causes were discovered, and the first steps were taken to try to put an end to this new epidemic. And the possibility of transmission to humans began to be mentioned.

The transmissibility of BSE and its membership in the family of prion diseases were the subjects of two brief notes published respectively in October and November 1988. When mice were inoculated with brain matter from cows with BSE, symptoms and brain lesions characteristic of scrapie appeared. And extracts of brain matter from cows with BSE contained fibrils similar to those found in animals with scrapie. These fibrils contained a protease-resistant protein whose amino acid sequence was very similar to that of the sheep prion. In all likelihood, then, this protein was the bovine prion.

In December 1988, John Wilesmith and colleagues published the first epidemiological study of BSE; it was based on nearly seven hundred cases and yielded data to which I shall return in due course. For the moment, I shall describe only its conclusions about the probable origin of the disease. Once a number of other hypotheses had been eliminated, the conclusion remained that the cattle had been infected by scrapie agent present in feed additives. Among those additives were meat and bone meals (MBMs) made from slaughterhouse and processing plant waste. Some of this waste was from sheep and could thus have been contaminated with the causative agent of scrapie, a disease that was still endemic in the United Kingdom. But, as we shall see, the use of animal meals to feed cattle was nothing new, and neither was scrapie. So why had BSE appeared only in 1985 and not before? The authors offered two main and not mutually exclusive hypotheses. The first hypothesis was that the very large post-1980 growth of the United Kingdom sheep population, the probable increase in the incidence of scrapie in that population, and the rising tendency to include the corpses of sick sheep in the waste from which food additives were made could have increased the quantity of scrapie agent in the material from which meal was prepared. The second hypothesis was that, owing to relatively recent changes in the methods by which MBMs were prepared, steps that had previously destroyed the scrapie agent might have been eliminated.

In spring 1988 the authorities were informed of the possible role of meat and bone meals in the emergence of BSE, and the British government enacted a series of measures in June and July of that year. Apart from the regulation making it obligatory to declare cases of BSE, the most important of these measures was a ban on feeding ruminants with feeds containing animal proteins derived from ruminants. If the theory advanced by Wilesmith and his colleagues was correct, that was supposed to lead to the gradual disappearance of BSE. Another series of measures taken in summer 1988 was intended to reassure consumers, who were worried about the possibility that since the scrapie agent had passed from sheep to cows, it might then pass from cows to humans. Chief among these measures was the obligatory slaughter and destruction of all animals suspected to have BSE, which of course posed the very difficult problem of how to incinerate the carcasses. Initially, farmers received compensation of 50 percent of the value of each animal that needed to be slaughtered. This was increased to 100 percent in February 1990 in order to encourage farmers to declare cases, which seems not to have been universally done. Other measures were adopted in 1988 and 1989 in light of what was known about the distribution of the scrapie agent in various sheep tissues. For example, because all cattle, even if healthy, could be in the disease's incubation phase, "at-risk" offal— brain, spinal cord, thymus, tonsils, intestines, and spleen—was withdrawn from the food chain beginning in November 1989.

After the end of 1988, while scientists waited to see the effect of measures taken to eliminate the epidemic, BSE was the subject of intensive study to better define the new disease, to test the theories of its origin, and to assess the risk of transmission to humans.

Unlike spongiform encephalopathies in sheep, goats, and humans, BSE is remarkably consistent both in its symptoms and in the nature and location of its nervous system lesions. In the vast majority of cases, the age at which the disease appears is between four and five years. The principal initial symptoms, as Wells and his colleagues had observed in the earliest cases, are nervousness, kicking, and problems with locomotion. As with all diseases of this kind, clinical development is insidious and inexorable. The animal's behavior begins to change—for example, it refuses to enter the milking barn and remains away from the rest of the herd or behaves aggressively toward them and sometimes toward the farmer. The most common locomotive symptom is a greater or lesser degree of incoordination in the hind legs; the animal frequently falls, and it rises initially on its forelegs, reflecting marked weakness in the hind legs. It is frequently hypersensitive to touching and to noise.

Obviously, these symptoms are very reminiscent of those observed in sheep with scrapie, but with one difference: In sheep, one of the most consistent symptoms is severe itching that leads to scratching by any available means and hence to the gradual destruction of the fleece. Nothing suggests that cattle suffer such itching. In general, the disease lasts for one to six months from the appearance of the first symptoms.

By 1990, the media were showing interest in an epizootic—the animal equivalent of an epidemic—that was growing to disturbing dimensions, and they took to calling BSE "mad cow disease." While catchy, that expression was only a partial description; a "mad" cow is primarily a nervous, fearful, hypersensitive animal that has difficulty in moving.

Of particular interest was epidemiological work with a view to identifying the origins of the disease. The first studies had blamed MBMs as the probable source of the initial contamination. The results of the ban on the use of such meals in ruminant feed demonstrated the accuracy of that theory; given the incubation period of diseases of this kind, the ban, which was enacted in July 1988, could not be expected to have an immediate effect. The number of declared cases thus continued at first to rise from a few hundred per month in 1988 to nearly three thousand per month in 1992. It then began to fall to no more than a thousand per month in 1995, and to about a hundred per month in 2000. It seemed very likely that MBMs had indeed caused the contamination. The gradual disappearance of BSE once these had been banned recalled the way that kuru vanished among the Fore after the prohibition of cannibalism. Not surprisingly, the public quickly came to the view that the appearance of BSE had been the result of turning cattle into carnivores by including meat and bone meal in their feed. This is partly true, but the matter is not that simple.

Using animal meals to feed cattle is not a recent practice. Note this extract from an 1893 work on the feeding of domestic animals:

Meat meal for fodder: the most concentrated food used in agriculture. These are the residua from the manufacture of Liebig meat extract. . . . Dairy cows and beef cattle, which are initially reluctant to eat it, soon come to accept it when it is taken in small quantities and thoroughly mixed with the rest of their feed—it is possible to go as high as 1.5 kilograms a day. Sheep are especially resistant, but also end up getting used to it.[1]

That last sentence, although written in passing, raises questions about the role that these meals might have played in the spread of scrapie among sheep.

We know, then, that the use of animal meals in cattle feed is not a new phenomenon and could not alone account for the emergence of BSE. Yet it is true that the use of these meals, especially for dairy cows, greatly increased after the 1960s and 1970s, with the development of intensive agriculture to maximize productivity. Including a small quantity of MBM in cattle feed met the animals' protein requirements. But again, for reasons of chronology, this increase in the use of such meals can probably not alone explain the emergence of BSE, which had to have occurred at least ten years earlier. Wilesmith and his colleagues finally offered a more likely explanation in March 1991.

At the time of their first study, they concluded that the BSE epidemic had resulted from simultaneous exposure of all English cattle during the period 1981–1982. Indeed, only cattle born during that time had been infected. The youngest came down with BSE at the age of two years, but the majority were four or five years of age. Thus, the origins of the disease could be traced to some event that took place in 1981 or 1982.

Wilesmith and his colleagues visited the forty-six British plants that until 1988 had converted a total of 1.3 million metric tons of meat and bone into meals to be used in animal feed. Without going into the rather unappetizing details of how this was done, we can say that the rendering process included steps to remove water and extract fats (tallow). These involved, first of all, a period of heating, which happened also to destroy the majority of microorganisms. Until the early 1970s, the material was heated in individual batches. Over two to three hours, these batches were heated to approximately 212 to 300 degrees Fahrenheit (approximately 100 to 150 degrees Celsius) and held at those temperatures for ten to twenty minutes. Beginning in 1971, batch processing was gradually replaced by continuous processing systems, which were more efficient and more economical. Perhaps these systems did not ensure that all the material being processed would reach a high enough temperature to inactivate the BSE agent. But here again, that explanation did not hold water. The changeover from batch to continuous processing took place very gradually between 1972 and 1984; in 1979 half of MBMs were manufactured using continuous processing, and that changed very little between 1979 and 1982. This is not consistent with a sudden appearance of the BSE agent in 1981 and 1982.

But another change, which did indeed coincide with the onset of contamination, was the abandonment of the use of organic solvents in tallow extraction and of the subsequent steam treatment at temperatures exceeding 212 degrees Fahrenheit (100 degrees Celsius). The change was made to save money, to streamline production, and to protect workers; in large part, it took place between 1980 and 1982. The concurrent dates obviously suggest that the outbreak of the BSE epidemic could have resulted from the abandonment of organic solvents in the manufacture of MBMs. The authors of the study offered the explanation that fats could have protected the BSE agent from destruction by heat but that extraction using solvents, by removing most of the fat, might have sensitized the agent to high-temperature steam treatment. The combination of solvent fat extraction and moist heat treatment could thus have been a very effective way to destroy the BSE agent in the manufacture of MBMs. The elimination of that process could have enabled the agent to survive and thus to find its way into the final product that was fed to animals.

In blaming MBMs, Wilesmith and his colleagues hypothesized that these had been contaminated by the scrapie agent, which had crossed the "species barrier" and had adapted itself to cattle. As we shall see, this was one likely hypothesis but by no means a certainty. Another hypothesis was that BSE was not a completely new disease, that there had been some sporadic cases in cattle, and that it was an agent originating in those cattle that was being spread by means of the meat and bone meal.

It was necessary first to know the prevalence of scrapie in the United Kingdom in the 1980s. Was it frequent enough that the scrapie agent could have caused massive contamination of animal waste used in MBM production? Answering that question was not as easy as it seemed. Indeed, gathering data on scrapie was no easier at the end of the twentieth century than it had been in the eighteenth or the nineteenth—and for the same reasons, as indicated in a study published in 1990, which stated: "The incidence and prevalence of scrapie in the United Kingdom sheep flock is unknown, and it is difficult to obtain information because of the potential economic loss to individual commercial and pedigree breeders of acknowledging the presence of scrapie in their flocks. An attempt was therefore made to obtain the information by using an anonymous self-administered questionnaire."[2]

The study concluded that about a third of British flocks had been affected by scrapie. In those flocks the incidence of the disease—that is, the number of new cases per year—was between 0.5 and 1 per 100 sheep per year. Those are large numbers. The total British sheep population was around forty million, so about 100,000 of them died of scrapie each year and had a good chance of winding up as an ingredient in meat and bone meal. There could be no doubt that this ingredient was liberally contaminated with the scrapie agent. Could it also have been contaminated by a BSE agent from a small number of cows that had spontaneously contracted the disease?

If Creutzfeldt-Jakob disease could sporadically appear spontaneously in humans, why should that not be true of BSE in cattle? That hypothesis was not broadly held by specialists who considered BSE to be a truly new disease that had never before been observed.[3] And many believed that a cow did not live long enough for the appearance of a sporadic disease whose human counterpart generally took at least fifty years to appear. Still, there is room for doubt, especially since the BSE agent, as we shall see, differs in many of its characteristics from the various strains of the scrapie agent that have been isolated from among sheep. But one thing seems certain: Contamination of cows by an infectious agent from other cattle did occur, but only in a second phase of the epidemic. Indeed, the incidence of BSE increased significantly beginning in 1989. Because the incubation period was three to five years, this increase certainly resulted from the recycling of the carcasses of the earliest victims of the disease, which had died between 1985, when the epidemic began, and 1988, when the use of MBM in cattle feed was banned.

One surprising aspect of the BSE epidemic was the distribution of cases within a herd. The number of cases per herd always remained very low; the initial herd, ten of whose cows were affected and which raised the original alarm for the authorities, was an exception. In the vast majority of cases, an affected herd would contain only a single sick cow. On average, the incidence of the disease was only about 2 percent in herds with at least one case. Given that animals of the same age in a given herd would have been similarly exposed to the infectious agent contained in the MBM, that distribution is somewhat surprising. Infection thus appeared to be random and unpredictable. This randomness did not seem to be the result of genetic differences between animals, but rather of the inefficiency of oral transmission, combined with the fact that the level of infectious material in the MBM might not have been very high. This recalls the situation of iatrogenic transmission of CJD via human growth hormone treatment. There too, while a large majority of batches of hormone had undoubtedly been contaminated, only a small proportion of children fell ill.

Once epidemiologists thought they understood the origin of BSE and its means of transmission, they turned to the question of how the epidemic might develop, taking into consideration the measures adopted by the British government starting in 1988.

In a 1991 study, Wilesmith and Wells put forward three hypotheses: First, the 1988 measures did put a complete halt to the transmission of the disease. Under that hypothesis, the number of cases would begin to decline in 1992 and the disease would disappear by 1999 or 2000. Second, the 1988 measures had halted oral transmission, but sick cows or those in the disease's incubation period could transmit BSE to their calves. That would not rule out the eradication of the disease, but would somewhat delay it until 2000 or 2001. And third, besides cow-to-calf transmission, some transmission could take place by direct or indirect contact between animals, as is the case with scrapie among sheep. Here, forecasts become very difficult, as the evolution of the epidemic would depend on how efficient such transmission was.

In fact, the number of cases began to decline in 1993, but the disease had not been eradicated by 2000. The most probable reason for this— one that Wilesmith and Wells did not foresee—is that the measures adopted by the British government in 1988 were not implemented with the necessary strictness. Either by mistake or by fraud, meat and bone meal continued to be used to feed cattle after 1988. When the authorities saw that the number of cases was not declining as quickly as expected, they enacted new provisions in 1996, making it a criminal offense to be in possession of MBM. As to cow-to-calf transmission, this proved to be possible, but its inefficiency (a maximum of 10 percent of calves born of a sick cow would be infected) meant that it had only a very limited impact on the evolution of the epidemic. Concerning animal-to-animal transmission, which could make BSE an endemic disease, like scrapie, there was nothing at that stage that suggested its existence.

BSE appeared first in the United Kingdom, where it has done the greatest damage, but it has also affected other countries to a lesser degree. While the total number of cattle affected by the disease from the beginning of the epidemic until the end of 2000 was nearly two hundred thousand in Great Britain, the figure was only about five hundred in Ireland and in Portugal, four hundred in Switzerland, and a little less than two hundred in France. The export of BSE from the United Kingdom seems easy enough to explain: It is linked to the export of meat and bone meal. The manufacturers of MBM, severely affected by the 1988 British government prohibitions, quickly sought overseas markets.

Exports to countries such as France consequently rose. The following year, to be sure, the French government prohibited MBM imports from Great Britain, and in 1990 banned the use of MBMs in ruminant feed. But here again, these measures were not enforced with the necessary strictness. The continued rise in the number of cases in France after 1991 through the year 2000 indicates that contamination was still occurring as recently as 1995 and 1996.

The first French case of *vache folle* (mad cow disease) was noted in February 1991. Then, and in all subsequent cases, the policy was to destroy not only the sick animal but the entire herd. This policy was intended to provide maximal protection for the consumer against any possible contamination by the BSE agent. It was based on the idea that all animals in a herd ran the risk of being in the incubation phase because they had probably eaten the same feed as the sick cow. The drawback here was that, in order to avert the destruction of a herd they had taken years to build up, some farmers might be tempted to conceal BSE cases on their farms.

22

FROM COWS TO HUMANS

THE POSSIBILITY THAT BSE could be transmitted to humans was considered as soon as the disease became known. In June 1988 the question was posed in a *British Medical Journal* editorial: "[We] are faced with the fact that spongiform encephalopathy, whether or not we are at risk from it ourselves, is now established in the cattle of this country. . . . There is no way of telling which cattle are infected until features develop, and if transmission has already occurred to man it might be years before affected individuals succumb."[1] And in September of that year an editorial in *The Lancet* raised the same issue. British experts then began to debate the probability that BSE could be transmitted to humans. This focused on three key areas: the relative infectiousness of various tissues from an infected animal, the efficiency of oral transmission, and the degree of protection that the "species barrier" offered to humans.

On the question of the relative infectiousness of various tissues, the experts looked to the precedent of scrapie, because previous studies had confirmed that the same tissues were infected in cases of BSE. Hence the prohibition of the use for human food of nervous system tissues as well as those of the digestive and lymphatic systems, which were the only ones known to be the site of significant infection in animals with scrapie. That prohibition affected all British cattle. There was no way to identify animals in the incubation phase, because they displayed no symptoms. The consumption of muscle tissue, on the other hand, was considered risk-free. That conclusion was probably correct, but could nonetheless be the subject of discussion. The various tissues, of course, were not as carefully sorted in a slaughterhouse as they would be in a research laboratory, and it is hard to rule out a certain amount of contamination of muscle tissue by nervous system tissue. The fact remains, however, that the extent of possible contamination of muscle tissue—that is, meat—was undoubtedly infinitesimal compared with that of at-risk tissues such as the brain.

The efficiency of oral transmission and the species barrier were addressed in 1989 by David Taylor in an article titled "Bovine Spongiform Encephalopathy and Human Health," in which he wrote: "Because bovine spongiform encephalopathy (BSE) has probably been caused by accidental transmission of the transmissible agent of sheep scrapie there is concern that humans may be at risk from BSE. Epidemiological and experimental evidence is examined which suggests that this is unlikely."[2]

That view was shared by many other experts and was not groundless. Even though oral transmission had been demonstrated in many cases, it had always proved to be very inefficient. Taylor pointed out that even for kuru it was not clear that contamination took place via the digestive system rather than occurring, through small skin lesions, in the course of handling the brain and viscera of individuals who had died of the disease. In a 1985 article, Prusiner's team had given a further example both of the possibility of oral transmission and of its inefficiency. In hamsters, oral transmission—in this case through cannibalism—required levels of the infectious agent a billion times higher than that needed for transmission by intracerebral injection. And the effectiveness of the species barrier proved to be variable: In some instances it could not be crossed, while in others it was breached fairly easily. The experts were heartened by the fact that BSE was viewed simply as scrapie that had been transmitted to cattle, and scrapie could not be orally transmitted to humans. We have seen that the hypothesis of such transmission had been considered to explain the higher frequency of Creutzfeldt-Jakob disease among Libyan Jews living in Israel, but that it had later been abandoned when the increased incidence of the disease was found to have a genetic basis. Moreover, a number of epidemiological studies had tried to establish a link between Creutzfeldt-Jakob disease and the consumption of meat from scrapie-infected sheep, but without success. If scrapie could not be transmitted to humans, why should BSE be any different?

There was little or no contamination of muscle tissue (meat); oral transmission was very inefficient; and the species barrier between sheep and humans, and hence supposedly between cattle and humans, was unbreachable—all serving to mitigate fears that BSE could be transmitted to humans. Yet these fears persisted. It had been known since Chandler's experiments in the early 1960s that the properties of the scrapie agent could change as the agent passed to a new host; for instance, the "drowsy" strain from a goat could become the "scratching" strain in a mouse. As a rule, once the species barrier had been crossed, the infectious agent would adapt itself to its new host—which implied a change in its properties. Furthermore, there was at least one almost certain example of a breach in the species barrier through oral transmission—precisely what was feared for the transmission of BSE to humans: There had been several scrapie epidemics on mink farms, where the mink had probably been fed meat from sick animals.[3] It had long been thought that these were scrapie-infected sheep, but it could have been cattle with spontaneous BSE or wild elks with a similar disease (chronic wasting disease). Remember too that in 1980 Gajdusek had reported oral transmission of kuru and of scrapie to squirrel monkeys, brought about by mixing contaminated material into the monkeys' feed.

These concerns were abruptly heightened in spring 1990, when spongiform encephalopathies were diagnosed in three domestic cats. Because this was apparently a new disease in cats, a connection was quickly made with BSE. The public came to the conclusion—without any real proof at that stage but, as it turned out, correctly—that the cats had been infected because their food had been contaminated with the BSE agent. This triggered a real panic among the traditionally staid British. If the BSE agent could be transmitted to cats through their food, why not to humans through theirs? And was there not a risk of cat-to-human transmission through simple contact?

So there was concern and there was vigilance. Had The Disease struck again? Every new case of Creutzfeldt-Jakob disease in Britain would be closely scrutinized. Would the incidence of CJD suddenly rise? Would it begin to affect populations that had generally been spared? Would it take hitherto undescribed forms? Or would it be limited, as in the past, to something less than one case per million, mainly affecting adults ranging in age from fifty to seventy, and taking one of the many forms in which it was already known?

The first warning came in March 1993. Creutzfeldt-Jakob disease was diagnosed in a British farmer in whose herd one cow had contracted BSE four years before. Could the man have been contaminated by the animal? The only possible route would have been through drinking the sick cow's milk, but neither the BSE nor the scrapie agent had ever been detected in milk. In light of data including the overall incidence of Creutzfeldt-Jakob disease, the large number of people who lived in contact with cattle, and so forth, this appeared to be a chance case unrelated to the BSE epidemic. The emergence of two further cases among farmers, one also in 1993 and the other in 1995, did not really give rise to concern. All three cases were of classic CJD, both clinically and in terms of their nervous system lesions.

The second warning came in April 1994. A fifteen-year-old British girl presented with a number of symptoms characteristic of Creutzfeldt-Jakob disease. Her youth was cause for attention. She had never had surgery or been given human growth hormone therapy, so iatrogenic contamination could be ruled out. DNA analysis showed no mutation in the prion gene, so this could not be inherited CJD. The patient was heterozygous at amino acid position 129, so she was inherently less prone to contract the disease. Did she, in fact, have Creutzfeldt-Jakob disease? The diagnosis was never confirmed.

The third warning came in October 1995. Two cases of "sporadic" Creutzfeldt-Jakob disease were detected, in a sixteen-year-old girl and in an eighteen-year-old boy. These were called "sporadic" because, as in the 1994 case, no specific iatrogenic or genetic cause could be found. But here, the diagnosis was confirmed, by biopsy in one case and by autopsy in the other. So this time the concern was palpable—and in fact one group of researchers concluded, "While the recent diagnoses of CJD in two teenagers in the UK may be coincidental and of no particular aetiological significance, they re-emphasise the need for continued epidemiological surveillance of CJD in the UK and in other countries."[4]

Then, the alarm was sounded in earnest in March and April 1996:

Ten cases of CJD have been identified in the UK in recent months with a new neuropathological profile. Other consistent features that are unusual include the young age of the cases, clinical findings, and the absence of the electroencephalogram features typical for CJD. Similar cases have not been identified in other countries in the European surveillance system.

These cases appear to represent a new variant of CJD, which may be unique to the UK. This raises the possibility that they are causally linked to BSE. Although this may be the most plausible explanation for this cluster of cases, a link with BSE cannot be confirmed on the basis of this evidence alone.[5]

Those few sentences come from the summary of an article published in *The Lancet* on April 6, 1996, by Robert Will and several other physicians, biologists, and epidemiologists working in the United Kingdom and other European countries. Its findings had been made public a couple of weeks earlier, on March 20, by the British Minister of Health. The possibility of transmission of BSE to humans in the form of Creutzfeldt-Jakob disease was no longer mere conjecture; it was now a likely hypothesis.

Although the BSE epidemic had already been known for ten years or so, this announcement marked the real beginning of the "mad cow crisis." It was a crisis whose effects would be enormous in many respects. Both in Britain and in countries that imported British meat, the public began to mistrust the whole range of products of bovine origin. Beef prices plummeted. On March 22, 1996, France declared an embargo on the import of beef and live cattle from the United Kingdom, an embargo that continues at the time of this writing. On March 27, the European Union placed a total embargo on all cattle and the products derived from them. Despite arguments put forward by France, that embargo was loosened in 1999. Overall, the mad cow crisis had serious social, economic, and political fallout.

What was so worrying about that article by Will and his colleagues? Certainly not any significant increase in the total number of British cases of Creutzfeldt-Jakob disease; the cases they described accounted for only ten of the 207 that had been studied since 1990, when the disease had become the object of epidemiological surveillance in Britain. Concern arose, first of all, because of the youth of the patients—an average age of twenty-nine. Among these, three individuals, including those earlier described in October 1995, had been younger than twenty when their symptoms first appeared. Concern arose also because of the unusual nature of those symptoms—among which psychiatric problems, especially depression, were prominent in the early stages of illness—and the absence of a characteristic electroencephalogram (EEG) pattern. In fact, none of these ten cases would have been thought a probable case of Creutzfeldt-Jakob disease solely on the basis of clinical data. Finally, these cases stood out by their neuropathological profile—by the appearance of their brain lesions. Although they all displayed the lesions that were typical of spongiform encephalopathies, they also showed amyloid plaques of a very particular appearance. These resembled the plaques found in the brains of members of the Fore tribe sick with kuru and were often surrounded by a "crown" of vacuoles that made them look like flowers, hence the name by which they were called: florid plaques.

Such plaques were nonexistent or extremely rare in cases of sporadic Creutzfeldt-Jakob disease, but had been described in the brains of animals with scrapie. Note too that none of these ten patients had undergone neurosurgery or human growth hormone therapy. In eight of them, the prion gene was examined; none of these displayed a mutation corresponding to one of the inherited forms of Creutzfeldt-Jakob disease. They were all homozygous at amino acid position 129, which was a risk factor for the disease. On the basis of all these criteria, it seemed that the ten young Britons had been affected by a new form of Creutzfeldt-Jakob disease, which was named "new variant Creutzfeldt-Jakob disease" (nvCJD).

It was The Disease in yet another new disguise.

From the very outset, it seemed likely that there was a causal link between BSE and the appearance of this new form of Creutzfeldt-Jakob disease. Specifically, nvCJD had appeared nowhere in Europe except the United Kingdom, and it had emerged between five and ten years after the period of the highest level of food contamination by the BSE agent at the end of the 1980s—that is, after a reasonable incubation period for diseases of this kind.

Reaction to these conclusions varied from panic to skepticism. It tended toward panic in Britain, where public awareness had been high since the outset of the BSE epidemic and which was on the front lines in confronting the new threat. And it tended toward skepticism in countries such as France, where, although many people had heard of the epidemic, they felt themselves insulated from it. In any event, while the link between BSE and nvCJD was probable, it remained to be proved scientifically. Many very convincing arguments would be put forward over the months and years to come.

The first came from French researchers who, more than two years before the existence of new variant Creutzfeldt-Jakob disease was known, had inoculated rhesus monkeys intracerebrally with extracts from the brains of BSE-affected cattle. When they displayed various neurological symptoms, the monkeys were sacrificed and their brains examined.

These contained florid plaques similar in every respect to those seen in new variant CJD patients. Such plaques had never been observed in rhesus monkeys inoculated in the same way with sporadic CJD. Since inoculating the monkeys with BSE caused the appearance of florid plaques, there was reason to think that the appearance of such plaques in humans also resulted from infection with the BSE agent.

The most convincing arguments were published in the United Kingdom in October 1997. They arose from the transmission of new variant Creutzfeldt-Jakob disease to mice. Overall, these experiments demonstrated that the nvCJD agent behaved like that of BSE, while differing from the agents of other forms of Creutzfeldt-Jakob disease. For example, while sporadic and iatrogenic CJD were extremely difficult to transmit to mice because of the species barrier, the new variant could be transmitted with relative ease, as could BSE. Conversely, an inbred line of transgenic mice was developed in which the mouse prion gene had been replaced by its human equivalent. This eliminated the species barrier, and it became comparatively easy to transmit sporadic or iatrogenic CJD. But it turned out to be difficult to transmit new variant Creutzfeldt-Jakob disease or BSE. Let us note in passing that the three cases of Creutzfeldt-Jakob disease that had been detected in British farmers in 1993 and 1995 developed like classic sporadic CJD, not the new variant, and were thus most likely to be unrelated to the BSE epidemic.

When added to the epidemiological arguments, the apparent fact that the BSE agent was the same as that of nvCJD led to the nearly inevitable conclusion that the latter human disease stemmed from the BSE epidemic, and more precisely from the consumption of meat or other products from contaminated cattle. Did this provide a basis on which to predict how the disease would spread within the British population and beyond? In August 1996, the epidemiologist Roy Anderson and a large group of other British researchers concluded their article on the development of the BSE epidemic with the following observation: "Whether or not the 12 reported cases of the new variant are the beginning of an epidemic remains uncertain, and will continue to be so for the next few years."[6]

In large part, that uncertainty persists today. As of early 2001, nearly a hundred people had contracted the disease, all of them in Britain apart from three in France and one in Ireland. In an August 2000 study, Anderson's team had projected the total number of cases that could occur in Britain as ranging anywhere from a hundred to 136,000. These projections were based on the development of the BSE epidemic and on the number of human cases since 1995, and took into consideration a number of hypotheses relating, especially, to the disease's incubation period in humans. The most pessimistic predictions (those in excess of ten thousand cases) all assumed an average incubation period greater than sixty years. At first glance that might seem unlikely, given the precedents for diseases of this kind. For kuru, to take one example, even though incubation periods exceeding forty years have been observed, the average has been only twelve years. If we accept an average incubation period of less than twenty years for new variant Creutzfeldt-Jakob disease, the number of victims, according to Anderson's projections, should not exceed a few hundred or a few thousand. But it must be recalled that, unlike with kuru, we are dealing here with contamination by an agent originating in another species, and that, in animal experiments, crossing the species barrier always involves longer incubation periods, sometimes three or four times longer. That is why the projections of Anderson and his colleagues, which contemplate a possible average incubation period of sixty years, are not altogether implausible. In any event, and however much we may want to know the extent of the tragedy in advance, we must be extremely careful as we consider these projections, which depend on a large number of parameters.

One frequently mentioned characteristic of new variant Creutzfeldt-Jakob disease is the youth of its victims. In principle, people of all ages have been exposed to the risk since the period 1985–1990, so all age groups should be equally affected by nvCJD. But nearly all patients have been between fifteen and forty-five years of age, and two-thirds have been under thirty-five. This could suggest either that young people had greater exposure than others or that they are more susceptible to infection—two hypotheses that are not mutually exclusive. One possible cause of greater exposure for young people could be contamination of those familiar little jars of baby food, as it appears that material from cattle brains might have been used in the manufacture of some of these foods. This could explain the very young age of some victims. As to greater susceptibility of the young to infection by scrapie, this had been observed among sheep—Thomas Comber observed it in 1772, and William Hadlow's team confirmed it in 1982. BSE too appears to be transmitted more readily to young calves, although it is impossible to say whether this is because of their greater susceptibility to the disease or because of feed containing more meat and bone meal.

To the extent that the transmissibility of BSE to humans has been proved, it is essential to put an end to the consumption of contaminated cattle products. Unfortunately, government measures put in place are insufficient to guarantee total safety. Even the systematic slaughter of affected herds cannot ensure that other herds' animals in the incubation phase will not be used for human consumption. Such animals should become increasingly rare because of the measures banning the use of meat and bone meal in feed, but they have not disappeared. Ideally, we would have diagnostic tests enabling us to identify animals in the incubation phase. But it must be recalled that at present there is no reliable diagnostic test for any form of spongiform encephalopathy in a living animal; the only tests are those carried out on the brains of dead animals. The most common of these involve looking for the protease-resistant, and hence infectious, form of the prion protein. Such tests can detect the infectious prion in sick animals or in those that are in the preclinical phase (that is, on the verge of displaying clinical symptoms). But they cannot detect the infectious prion in animals at earlier stages of incubation, when they could nonetheless be possible sources of contamination. Current efforts by researchers to increase the sensitivity of these tests could soon solve this problem.

23

FROM COWS TO SHEEP?
FROM HUMANS TO HUMANS?

AS THOUGH FEARS ABOUT EATING BEEF and other cattle products were not enough, some people began to worry about the possible risks of eating products from other animals—first and foremost sheep. The fact is that cows were not the only animals to have been given potentially contaminated animal-based feeds; these had been fed also to sheep, hogs, chickens, and even fish. Could it not be risky to eat meat or other products from those animals as well—especially since, for some of them, animal-based meals had still been a part of their diet until recently? Let us look first at the case of sheep, which pose the greatest risk and which also are particularly interesting from the scientific standpoint.

Until recently it had been accepted that BSE had come about through the transmission of scrapie to cattle via meat and bone meal (MBM) in their feed. In that scenario, reverse contamination of sheep through the consumption of MBM made from BSE-infected cattle would have been a simple case of "return to sender." Such sheep ought simply to have contracted scrapie, which is known not to be communicable to humans. Then why should we worry about possible contamination of sheep by the BSE agent? Because the properties of the latter are very different from any of the various strains of the scrapie agent that have been identified to date.

The characteristics specific to the BSE agent, compared with those of the various strains of scrapie, are manifested, among other ways, in their different effects on different inbred lines of mice, in terms of such factors as incubation period and the areas of the brain that are principally affected. There has been speculation about the origin of this difference between the BSE agent and the known strains of the scrapie agent. Some scientists think that the scrapie agent could have been modified during the processing and manufacture of MBMs. Others suggest that a particular strain of the scrapie agent, perhaps very rare, and thus undetected until then, might have surfaced because of its unusual resistance to processing or its ability to infect cattle orally—unless, of course, the BSE agent does not come from sheep at all, but rather from those few cattle that contract the disease spontaneously, in which case the agent would be of bovine origin. If the BSE agent came from sheep, and if those sheep had been fed MBMs, we would have expected a scrapie epidemic to arise among sheep at the same time as the BSE epidemic broke out among cattle. This would have been even more likely because the infectious agent would not have had a species barrier to cross. But the fact is that the 1980s saw no scrapie epidemic among sheep. Does that not suggest that the BSE agent did not originate in sheep, as had long been believed, but in cattle?

Whatever the reason why the BSE agent differs from known strains of the scrapie agent, the most worrying fact is that it retains its properties when it passes to different animals, including sheep. In theoretical terms, that stability cannot fail to be of concern. It would seem to contradict the outcome of earlier experiments on transmission between species. Recall, for example, that when scrapie was induced in hamsters using an agent from mice, the result was an infectious agent with the characteristics of a hamster agent, not a mouse agent. And that stability is worrying from a public health standpoint: It means that products originating in any animal infected by the BSE agent can be a potential source of human contamination, even if the animal is a sheep. It is encouraging that the BSE epidemic among cattle has not been followed by a detectable increase in the incidence of scrapie in sheep flocks; if sheep have been contaminated by the BSE agent, at least it appears to have remained limited. But the matter deserves careful follow-up. Although BSE does not seem to be transmissible from cow to cow, there is possible reason to fear that if it were transmitted to sheep it could then be transmitted within the flock, as scrapie is. It could then become entrenched forever. In 1996, as a preventive measure, the French government prohibited the sale of meat from sheep with scrapie.

And what about the other animals commonly eaten by humans? Supposing that the BSE agent can develop in them—which is by no means proved—it would seem that none of them live long enough for the infectious agent to multiply to a significant level. BSE has been transmitted to pigs, but only with great difficulty. Oral transmission has not been achieved, only transmission by intracerebral injection of extremely large quantities of the infectious agent. Moreover, the earliest symptoms took more than a year to appear. In the light of that outcome, it seems very unlikely that hogs could be contaminated by the consumption of MBMs containing the BSE agent. Moreover, hogs raised for meat are slaughtered before they reach six months of age—too soon for the infectious agent to multiply significantly.

None of the various attempts to transmit BSE to chickens has succeeded. And as for fish, it seems unlikely that the BSE agent could multiply in them, given the biological distance between fish and cattle—here, the species barrier ought to be truly unbreachable. And let us note that, again as a preventive measure, the French government in 1996 prohibited manufacturers of MBM intended to feed animals other than ruminants from using slaughterhouse waste from animals not declared fit for human consumption. A provisional total ban on MBMs was adopted in late 2000, not so much because of any risk to hogs, chickens, or fish, but because they could be used accidentally or fraudulently to feed ruminants.

A September 2000 article in *The Lancet* raised a new question about new variant CJD: Could it be transmitted by blood transfusion? This is an important question. If such transmission were possible, and if there were a large number of individuals in the incubation period of new variant CJD, we could be facing another contaminated blood tragedy. In an editorial accompanying the article, Paul Brown, a key expert in this field, summed up what was known concerning the issue of contaminated blood with the following three points:

· Blood (especially white blood cells) from animals experimentally infected with the scrapie agent or Creutzfeldt-Jakob disease was infectious when injected intracerebrally or intraperitoneally into animals of the same species.

· Among animals infected naturally, including BSE-infected cattle, all attempts to show that their blood was infectious had failed.

· Epidemiological data had not disclosed a single case of CJD attributable to transmission by blood transfusion or the administration of blood derivatives, either among hemophiliacs or among other patients who had repeatedly been given blood derivatives.

That would seem to be fairly reassuring. Blood does not appear to be infectious except in highly artificial situations when encephalopathies are transmitted in laboratory conditions. But none of these data relate to new variant Creutzfeldt-Jakob disease, which in so many ways displays very distinctive characteristics. Unable to conduct human experiments, the authors of the September 2000 article reported the preliminary outcome of an experiment on the transmission of BSE among sheep. Nineteen animals were given the BSE agent orally, comparable to the way humans would be infected through their food. At various stages of the incubation period, blood was transfused from these sheep to others presumed to be free of scrapie or BSE. As of the article's publication date, one transfusion seemed to have caused transmission of the disease. The donor sheep, from which blood had been drawn 318 days after oral inoculation, contracted the disease on the 619th day. The blood had thus been drawn about halfway through the incubation period. The recipient displayed the first symptoms of encephalopathy 610 days after the transfusion. This outcome—which obviously needs to be reproducible—shows that the blood of an animal in the incubation phase of BSE can be infectious. And so, by extension, could the blood of a human in the incubation phase of new variant Creutzfeldt-Jakob disease.

What could health officials in countries affected by new variant CJD, as well as elsewhere, do in the face of this threat—a threat very difficult to assess? In France, everyone who had received injections of hormones extracted from human pituitary glands—and thus at greater risk of contracting Creutzfeldt-Jakob disease—was already forbidden to donate blood. But it was impossible to extend this to everyone who might possibly be in the incubation phase of new variant Creutzfeldt-Jakob disease, because no one knew who they were. On the other hand, some countries unaffected by the disease refused to accept blood donations from people who had lived in countries viewed as being at risk. In France, a complete halt to the use of blood from within the country would surely have had serious public health consequences, so a different approach was taken: not to accept blood donations from anyone who had received a blood transfusion. This would not prevent possible contamination by transfusion from individuals themselves contaminated by oral transmission, but it would prevent the recipient of a transfusion from in turn contaminating one or more other people. Thus, given what was likely a very small risk, it can be said that all reasonable protective measures were taken.

Other tissues from people who were in the incubation phase of nvCJD, apart from blood, could also be a source of contamination. It appeared that, unlike that of classic CJD, the nvCJD agent was widely distributed within the organs of the lymphatic system, perhaps because of the route by which it entered the body. This could require changes in the procedures for sterilizing surgical instruments, similar to those already in place for neurosurgery.

Many scientists were also wondering about the risk of mother-to-child transmission. At present, we have no data on such transmission of new variant CJD. But data on kuru are rather encouraging. No child born after the cessation of cannibalism has contracted kuru, even though at least a hundred of them were born of mothers affected by the disease. This enables us to hope that mother-to-child transmission of new variant Creutzfeldt-Jakob disease will prove similarly nonexistent or very rare.

Considering all the effects of the mad cow crisis and what is known about its probable origin—changes in the manufacturing processes of meat and bone meal intended for animal feed—it is surprising that the crisis broke out in the United Kingdom. Of all countries, it was Britain, with its long history of scrapie, that had the most advanced knowledge of that disease. As we have seen, the work of British veterinarians and veterinary scientists had been crucial. And it appears that at least one veterinarian had indeed advised against the proposed changes in the processing of meat and bone meals, but that advice was not heeded. If any country had reason to be concerned about the possible dietary transmission of spongiform encephalopathies, it was Britain.

24

THE SECRET IN THE CLOSET

WE HAVE BEEN ON THE TRAIL of The Disease for three centuries now. If it has managed to evade capture for so long, this is due in large part to its disturbing ability to alter its appearance. The closet in which this criminal stores its disguises seems to be infinitely deep. So what is its secret? Can prion theory help us figure it out?

From the very outset we knew that The Disease could change its appearance. Think of how hard it was to realize that scrapie in all its forms was in fact a single disease—hence, the many names it was given. We see this again in the varied forms that human spongiform encephalopathies can take: classic Creutzfeldt-Jakob disease, kuru, iatrogenic Creutzfeldt-Jakob disease, new variant Creutzfeldt-Jakob disease, Gerstmann-Sträussler syndrome, and fatal familial insomnia. We have mentioned it also in connection with scrapie experimentally induced in mice—a subject worth briefly returning to, because it is in that form that the variability of the disease has received the most study and has been "labeled," so to speak, through the half-century of long-term research carried out by Edinburgh veterinary scientists including Alan Dickinson, Richard Kimberlin, Moira Bruce, and a number of others. A

score or so strains of the scrapie agent have been identified; these origi-
nated in various scrapie-infected sheep and were then sequentially
transferred to different inbred lines of mice. These strains differ mainly
in the disease's incubation period in four lines of mice and in the location of the lesions they cause in the animals' brains. The characteristics
of these prion strains are stable and remain unchanged through successive passages within a single mouse inbred line. There is, in fact, an astonishing degree of stability. For example, for a given quantity of agent
administered intracerebrally, the incubation period for a given strain
can be predicted to within three or four days. On the other hand, incubation periods for two different strains can vary from 150 to 600 days.

Besides this variety of appearances within a given species, there are
the changes that have been observed when the agent is transmitted from
one species to another and thus crosses the species barrier. It then ac-
quires new properties including, but not limited to, adaptation to its
new host and becoming efficiently transmissible to other individuals of
the same species. What is the source of the agent's ability to differentiate
itself in various strains, each with its own distinct properties?

If we were dealing with a conventional infectious agent, such as a
bacterium or a virus, this would present no particular problem. The
properties of such agents are determined by their genes, so they can be
modified by mutations corresponding to accidental changes in the sequence of bases in their nucleic acid. For example, that is how such conventional infectious agents adapt to new hosts: Mutants that develop
more rapidly than the original strain quickly take over. But, according
to prion theory, the infectious agent of a transmissible spongiform encephalopathy (TSE) is without nucleic acid. It is a protein that can exist
in two forms, the "normal" form and the pathological "infectious" form.
How can a single pathological form be consistent with the disease's
many and varied clinical pictures? Many scientists believe that this variability could be explained only if the agent contained genetic material—
nucleic acid that no analysis has been able to detect and that could be
subject to mutation. But a different explanation has been put forward

by proponents of prion theory. It is somewhat complex, but it is based on convincing experimental findings.

This explanation, which emerged between 1995 and 2000, is that the variability of the agent of TSEs has a twofold origin: genetic and nongenetic.

The genetic origin lies in the amino acid sequence in the prion protein, variations in which can have a variety of effects. In cases of inherited Creutzfeldt-Jakob disease resulting from mutations in the prion gene, the disease appears spontaneously with higher frequency than among the rest of the population and with symptoms that vary depending on the mutations present in the gene. Still, in humans, the nature of the position-129 amino acid is known to affect susceptibility to infection by a pathogenic prion; but it has recently been shown also to affect the nature of the nervous system lesions. In mice, prion gene mutations have been seen to affect both the incubation period and the site of lesions within the brain. And among sheep, the amino acid sequence of the prion gene determines incubation period and susceptibility to infection. In all these cases, it is clear that differences in the prion's amino acid sequence, which result from differences in the gene's base sequence, lead to differences in the conformation of the protein and hence to differences in pathological effects.

While the genetic origins of prion variability are entirely conventional, its nongenetic side is far less so, for it suggests that there are not one but several possible infectious conformations for a prion protein of a given sequence, and that each of these can impose its conformation on a normal protein as it bestows the "kiss of death." For most biologists such a hypothesis long seemed the most far-fetched of fantasies. Since the founding of molecular biology, the prevailing thinking had been that the final structure of a protein was completely determined by its amino acid sequence. It was thought that a protein's conformation was predetermined and was only susceptible to minor changes through the effects of an outside agent—in enzyme activation and inhibition, for example. So, is the prion protein an exception to the general rule? Can it exist in several different conformations, all of them stable and each causing different kinds of nervous system lesions? How could such a notion fail to be met with incredulity? Yet experimental results supported this astonishing hypothesis.

The first such support came not from one of the major laboratories engaged in the study of scrapie, BSE, or human encephalopathies, but from a Wisconsin lab that since the 1960s had been studying the scrapie that mink had contracted from their feed. The results of that laboratory's study of two strains of infectious agent that it had identified were then extended to transmissible spongiform encephalopathies in sheep, cattle, and humans. These studies used biochemical methods to show that a single agent could actually exist in several conformations.[1] Among the strains adapted to hamsters, Prusiner's team succeeded in identifying as many as eight different ones. Since various conformations of the protein exist, they can, one way or another, be seen. And an infectious protein of a given conformation seemed to be able to impose that conformation on a normal protein.

Experiments to demonstrate the latter phenomenon were aimed at producing the "kiss of death" in a test tube. The theory predicted that one could make a normal protein protease-resistant and infectious by mixing it with an infectious protein. Could this be done? Attempts long remained unsuccessful—until researchers had the idea of partly "unspooling" the normal protein by treating it with the proper chemical before mixing it with the infectious protein. Then, in the presence of the infectious protein, the original protein would "re-spool" in a protease-resistant form. If infectious proteins with a variety of conformations were each mixed into separate batches of partly unspooled normal proteins, these would re-spool, having adopted the conformation of the infectious protein that had been added. These experiments seemed to show that, consistent with the theory, the infectious protein could impose its conformation on a normal protein. Unfortunately, though, experimental conditions made it impossible to know whether the protein that had become protease-resistant had also become infectious.[2]

The prion's variability thus resulted from the sequence of its constituent amino acids (its genetic origin), and from the three-dimensional conformation transmitted to it during the "kiss of death" (its nongenetic origin). Working on that basis, a group of researchers sought to correlate the varied clinical pictures of sporadic Creutzfeldt-Jakob disease—which had in the past given rise to as many as sixteen different names for the disease—with the nature of the prion protein's position-129 amino acid and with the conformation of the protein, as analyzed using biochemical techniques. They studied nineteen patients, grouping them into four homogeneous categories on the basis of clinical symptoms and genetic and nongenetic criteria. The variability of sporadic Creutzfeldt-Jakob disease could thus be the result of the various possible conformations of the protein.

Another worrying aspect of prion biology relates to the species barrier and how the prion can cross the species barrier by changing its properties. The prevailing thinking is that the species barrier exists because the prion's proteins prefer to associate with other proteins that have the same amino acid sequence, and hence with proteins from the same species. Why should this be so? That question leads us to a key element of prion theory, but one on which, oddly, not a great deal of attention has been focused to date. What is the cause of the "kiss of death"? Why does the infectious protein associate with a normal protein and make it infectious as well? For two proteins to associate, their surfaces must display complementary regions that enable them to remain bound together. In biochemical terms, they must have an affinity for each other. This cannot happen with just any two protein species; for the "kiss of death" to take place, the normal protein and the infectious protein must be able to associate thanks to complementary regions, which do not occur at random. What could be the origin of these complementary regions? Two hypotheses come to mind.

Under the first, the prion protein would normally exist in the form of a dimer or an oligomer;[3] its surface would thus contain complementary regions conducive to association between identical molecules. In the dimer's or oligomer's configuration, an infectious-protein molecule would replace a normal protein, hence the "kiss of death." Such an association would be more difficult—but still possible—if there were differences in the sequences of the infectious and the normal protein, as there would be if the infectious protein came from a different species. Here, the complementarity between the regions that should associate could be imperfect. Under this hypothesis the normal protein would be either a dimer or an oligomer, and it would be necessary to find out whether this is the case.

Under the second hypothesis, the normal protein would be a monomer, and the regions where it associated with the infectious protein would typically be within the protein and would help maintain its three-dimensional structure.[4] Under normal circumstances, these regions would be buried in the protein, but they could be exposed in the course of the protein's temporary partial unspooling; this would make possible an association with the infectious protein's complementary regions, which would likewise be temporarily exposed. Here again, such association would be easier between identical proteins than between proteins from different species. This "unnatural" dimerization of a normal protein and an infectious protein would create access to associative regions that would normally be concealed within the molecule, and these would then be able in turn to trigger other normal protein molecules, beginning a process of polymerization. Perhaps this polymerization involved the emergence of the protease resistance that is characteristic of the infectious form.

In both hypotheses the "kiss of death" is made possible by the presence of preexisting regions of interaction in the normal protein—on the surface according to the first hypothesis and internally according to the second. This may all be plausible, but we must not forget that it is based on the truly heretical idea that a protein can exist in a number of stable conformations corresponding to different strains—a phenomenon that, if true, would be nearly unprecedented. That is why many scientists remain doubtful, thinking that, even if a protein can exist in several conformations, "something"—perhaps a nucleic acid—must be imposing these conformational changes on the prion. The fact is that we still lack ironclad proof of the prion hypothesis.

How can we get such proof? The ideal experiment might be as follows: A normal protein would be prepared through genetic engineering, from a bacterium or a yeast into which a prion gene had been introduced. Better still, it could be chemically synthesized from amino acids. A small quantity of infectious protein, purified as completely as possible, would be added to this normal protein. Then we would need to find conditions under which the infectious power of the mixture increased significantly. Another solution of normal protein prepared in the same way would then have to be made infectious by the test tube–synthesized infectious protein, and so on. Such an experiment would continue until, owing to the successive dilutions, no more of the initial infectious protein remained in the mixture. By thus creating successive "cultures" of the prion, we would be repeating the experiments by which Pasteur proved that the anthrax bacillus was indeed responsible for that disease. Unfortunately, no one has yet been able to demonstrate the development of an infectious prion in a test tube.[5]

Although doubts remain about the validity of the prion hypothesis, let us take it as established. Can we then assume that the hunt is over, that The Disease has been unmasked? To be sure, there has been considerable progress, but many gray areas remain. Furthermore, to identify The Disease is not necessarily to conquer it. It continues to claim victims. Entire nations—Britain, France, and others—live in fear. Not only young people treated fifteen or twenty years ago with human growth hormone but, because of the BSE epidemic, everyone wonders how many people will be affected by Creutzfeldt-Jakob disease. And everyone fears becoming a victim.

25

UNMASKING "THE DISEASE"

SO, THE DISEASE APPEARS to be caused by a molecule, a protein that is more or less the same in all species that fall victim to it. Does this hypothesis in its present form account for all the various characteristics of The Disease? Before saying that it does, we should look at a few more questions.

One obvious question, which we addressed earlier, relates to the precise nature of the changes in conformation that turn the prion protein infectious. The conformation of the normal protein has been described, but not those of all the various infectious forms, whose insolubility makes that a very difficult task. And the mechanism of the "kiss of death" remains to be described as well. Does it occur specifically between a normal-protein monomer and an infectious-protein monomer, or between two dimers, or even—in a process reminiscent of crystallization—between a normal-protein monomer or dimer and an infectious-protein polymer to form the beginnings of a fibril?

Another question is: Where exactly does the "kiss of death" take place? Experiments on cultures of prion-infected cells show that a protein that will later become infectious is first exported as a normal protein to the surface of the cell in which it had been synthesized, and that its conversion into an infectious protein occurs when it is reinternalized during the normal process of renewing the cellular membrane. It could be that the association between the normal and infectious proteins occurs at the surface of the cell; that, simultaneously with the membrane, the complex is internalized into intracellular compartments known as endosomes; and that the conformational change that makes the normal protein protease-resistant takes place in these compartments. This resistance enables it to survive and build up in the cell. This hypothetical scenario, while plausible, has yet to be confirmed.

Then we face this key question: Is the accumulation of infectious protein in cells—in this case neurons—responsible for cell death and thus for the appearance of the vacuoles that are characteristic of the disease? For all its importance, this question has yet to be satisfactorily answered. Although a buildup of protease-resistant protein in a cell seems capable of causing serious changes in the cell's functioning, it remains unclear exactly why this should bring about the cell's destruction. Moreover, there are documented cases of cell death without any detectable buildup of protease-resistant prion protein. In short, we do not yet know how the prion kills.

Whatever the process may be that causes prions to destroy neurons, the clinical symptoms depend on which neurons are under attack. That is a general rule of nervous system diseases—for example, the destruction of certain neurons in the cerebellum causes balance disorders, while the destruction of neurons in the cerebral cortex can cause psychiatric symptoms. As we have seen, the differing symptoms of transmissible spongiform encephalopathies—the various disguises adopted by The Disease—could depend on the way the prion enters the body or on its conformation, which itself is determined by both genetic and nongenetic factors. This raises another question: How can the prion's conformation determine which cells it will attack? To date, we have no answer to that question.

We just referred to the way the prion enters the body. Intracerebral injection, which is used in animal experiments and which has occurred accidentally in humans during neurosurgery using contaminated instruments, places the prion in direct contact with its target cells. But that is not the case with iatrogenic infection resulting from injections of human growth hormone or with the oral infection that is probably responsible for the persistence of scrapie in sheep, the contamination of cattle and humans with BSE, and the kuru epidemic among the Fore people. So how does the prion find its way to the central nervous system, to the brain?

The question of oral contamination raises a problem of some seriousness. The normal fate of an ingested protein is to be broken down by proteases in the digestive tract and thus to be a source of amino acids for the human or animal that ingested it. And a protein that is not broken down is, in principle, eliminated with other digestive waste. An undigested protein is not supposed to pass through the intestinal barrier. The only well-documented exceptions are bacteria-produced toxins that have very special molecular properties that enable them to recognize receptors on the surfaces of certain intestinal cells and then to penetrate those cells. So do prions act like toxins? No one knows for sure, but we must bear in mind that the prion is unquestionably far less able than a toxin to pass through the intestinal barrier. Perhaps this slow, ineffective passage comes about through the infectious prion's interaction with cells on the surface of the intestines whose membranes possess a normal prion protein. The infectious prion could reproduce within such a cell and, perhaps after it has destroyed it, could be transmitted to another cell possessing the normal protein, and so on. That is nothing more than a theory, but it brings us to the question of the nature of the cells that make possible the prion's journey from the intestines to the brain.

One thing seems certain: The infectious prion is not simply transported intact to its target cell. It must reproduce along the way. In a sense, it is the descendants of the original molecules that reach the brain. The prion spreads by means of cells that synthesize the normal protein, which along the way is converted into the infectious form. What kinds of cells are involved? A fair body of data suggests that immune system cells probably play a role. Recall that, after the intestine, lymphatic ganglia were among the first organs to be infected in the course of natural scrapie in sheep. Other immune system organs, especially the spleen, seem also to be major reservoirs of prions during infection. The first stages in the prion's journey, including its passage through the intestinal barrier, seem to occur by means of the immune system; the peripheral nervous system (that is, the nerves) then takes over. Direct transmission via nerve endings in the intestinal wall also seems possible.

It is obvious that the prion theory leaves many questions unanswered. But by and large it is convincing, so convincing that it is accepted by a great majority of specialists in the field. It would be tempting to conclude that the hunt is over, that The Disease has been unmasked. But has it been conquered?

26

HAVE WE CONQUERED "THE DISEASE"?

IN NOVEMBER 2000 a French television network broadcast a program titled "Mad Cow: Running Scared." The Disease—which had been lurking in the days of Louis XV and which we've been hunting down for the past three centuries—is still spreading fear today. Even if we have unmasked it, we have not conquered it. The public has been deeply shocked by images of young people, of mothers, unable to move about, speak, or even show any sign that they understand what is said to them, incontinent and awaiting a certain death. People are suspicious of the food they buy. The anxiety of people who are afraid that they or their children will be struck down by this terrible disease and the concern of farmers and manufacturers who are threatened by bankruptcy and ruin have had enormous political repercussions.

What lies in the future for the mad cow crisis and its consequences?

Because of the measures taken by the United Kingdom and other European countries to stop the use of meat and bone meal in animal feed, it seems likely that the BSE epidemic will come to an end by 2004 or 2005. Of course, uncertainty remains about transmission among cattle, which could somewhat delay the elimination of the disease. But the way the disease has evolved in the United Kingdom to date suggests that such transmission is ineffective if it exists at all.

Unfortunately, the hypothesis that BSE can be transmitted to sheep cannot be ruled out, although if this has taken place it cannot have been on a large scale. But even if it is limited, it could make BSE endemic among the sheep population, as scrapie is; it would then be very difficult to eradicate. Fortunately, ongoing research in Britain has not yet found the BSE agent to be in sheep.

If, as we believe, cattle and sheep products have already ceased (or are quickly ceasing) to be a source of infection, the epidemic of new variant Creutzfeldt-Jakob disease in humans should come to an end. But when? As we have seen, that will depend on the disease's incubation period. There is no doubt that the United Kingdom will be the country hardest hit, but France and other countries that had indigenous BSE cases of their own or that imported large quantities of British cattle products, especially in the late 1980s and early 1990s, will also be affected. The worst case, according to Roy Anderson and his colleagues, is that about 130,000 people in Britain could be affected in the coming decades. It is harder to make predictions for France, where only three cases have been reported to date (see Chapter 27). It is probable that most current or future cases are, or will be, the result of eating cattle products imported from the United Kingdom in the late 1980s or early 1990s. Since exports to France accounted for about 5 percent of total British output, we can estimate that the number of French cases will likewise be about 5 percent of the number of British cases, or a maximum of six to seven thousand. But these estimates, which conflate many different hypotheses, rest on very shaky scientific ground.

Finally, the greater the number of primary victims contaminated through their food, the greater the risk of secondary victims contaminated through transfusions or surgical procedure. That risk is impossible to assess; indeed, it is entirely theoretical and may even be nonexistent.

In the meantime, as we wait for BSE to be eradicated, let us return to a question we posed in the Prologue: What can we do to protect ourselves? Milk and other dairy products appear to be risk-free; neither the BSE agent nor the scrapie agent has been found in them. At present, eating French beef seems to pose very little risk. For one thing, judging by recent tests carried out by the French food safety agency, the number of animals that could carry the BSE agent seems to be fewer than two per thousand.[1] And, most important, muscle tissue—meat—contains no detectable quantity of the infectious agent. On the other hand, we must take care (as we have done for some time now) about eating "at-risk offal" such as the brain and the spinal cord, which should not be marketed.

Is it possible to keep cattle or sheep from contracting BSE or scrapie? A theoretical approach has been suggested by the fact that removing the prion gene has been shown to make mice immune to scrapie. Would it not be possible similarly to remove the prion gene from our livestock? The idea is by no means absurd, but—assuming it is possible—it could not immediately be put into effect. The central question relates to the short-term or long-term effects of removing the prion gene from those species, which could have totally unforeseeable consequences, including posing a new threat to the consumer. It is not even certain that cows or sheep lacking the prion gene would be viable, as mice proved to be. If it proved possible in those species to remove this gene without harm, much work would still need to be done to do so in a sufficiently large number of individual animals to retain genetic diversity. This approach is thus not suitable for ending the BSE epidemic among cattle—which we have every reason to hope will quickly come to an end on its own. But it might be worth considering if by bad fortune BSE were to become a permanent endemic disease among sheep.

What hopes do we have of diagnostic tests for BSE? Present tests, which detect the presence of protease-resistant prion protein in an animal's brain, need quickly to be made more sensitive. If systematically used on animals after slaughter, such tests could make it possible to detect not only animals in the clinical stage of the disease but also those in the incubation period. But it will always be difficult to know whether a test is sensitive enough to catch every prion-carrying animal in the earliest stages of incubation, whose meat and offal could be a source of contamination. In addition, farmers obviously want to know if their animals are contaminated before sending them to the slaughterhouse. In other words, they want a test that can be carried out on a living animal. No such test exists at the moment, because the prion protein seems to be absent, or to exist in infinitesimally small quantities, in the fluids that can be taken from a living animal without harming it, such as blood and urine. This is no less true for humans, for whom the only conclusive means of diagnosis, short of a somewhat risky and complicated brain biopsy, continues to be postmortem examination of the brain.

If we cannot completely avoid all risk of BSE contamination in humans—at least, as long as the epidemic persists in meat animals—would it be possible to prevent Creutzfeldt-Jakob disease, or even to cure it? This question relates not only to new variant CJD but also to the other spontaneous, inherited, and iatrogenic forms. In principle, it is difficult to conceive of a vaccine; assuming it were possible to trigger an immune reaction against the prion protein, this could give rise to an autoimmune disease that would use the immune system to destroy cells that had the normal protein on their surface. It might be possible to try to prevent the synthesis of the normal protein in infected individuals—but, apart from the fact that there is no obvious way of doing this, we come back to the question of the protein's function. It is by no means certain that, in humans, the lack of this protein would be harmless. The ideal weapon would be one that prevented the normal protein from converting into the infectious protein. Experiments along that line are now under way. There have been some promising results, but there is a long way to go before these results yield a treatment.

27

2001

AS THE FRENCH EDITION of this book went to press toward the end of 2000, Europe's mad cow crisis was at its height. A year later, the crisis was still with us but seemed less intense. Beef consumption had picked up somewhat in France, but concerns about lamb and mutton emerged. In countries that had been BSE-free, the first cases of the disease triggered panics like those that had been experienced in the United Kingdom and then in France. In the scientific arena, many—and often extremely interesting—findings were announced, but their impact remained uncertain.

DIAGNOSIS OF TRANSMISSIBLE SPONGIFORM ENCEPHALOPATHIES (TSEs)

One particularly active area of research was diagnosis. Traditional tests were able to detect the prion in tissue sections or brain extracts. They relied on antibodies that could recognize the prion protein in both its normal and pathogenic forms; the two could be distinguished because the normal form was destroyed by proteases under conditions when the pathogenic form was not. These tests have two main limitations: Their sensitivity is limited, and they can be carried out only after death. Researchers have therefore focused on developing more sensitive tests that can detect small quantities of prion in animals or humans during the disease's incubation period, preferably in bodily fluids that can be drawn from a living subject.

In late December 2000, Adriano Aguzzi's Zurich-based team reported that a blood protein, plasminogen, bound to the pathogenic prion in mice without binding to the normal form of the prion.[1] The explanation for this is not clear, but the observation opened the way for a future diagnostic test that would use plasminogen as a specific reagent and that could be developed on a commercial scale.

Another promising outcome was published by an Israeli laboratory in August 2001. The article reported the detection of a form of prion in the urine of animals and humans infected with TSEs.[2] This protease-resistant form of prion was detected by traditional methods after concentration using a high-speed centrifuge. Had this lab found the Holy Grail? Had it succeeded in detecting prion in the most easily obtained of all bodily fluids, and doing so well before the appearance of any symptoms? Their outcome has yet to be confirmed in other laboratories.

The most fascinating finding in the diagnostic realm—fascinating not only in practical terms but also because of its theoretical implications—was published in June 2001 by Claudio Soto's team at the Serono corporation's Geneva laboratories.[3] Their article describes a simple technique to increase the sensitivity of the test to detect the pathogenic prion by a factor of between ten and a hundred. We know that the normal, protease-sensitive prion can be converted into a protease-resistant form through incubation with the pathogenic prion. But this conversion process is not particularly effective and requires a large quantity of the pathogenic prion. The authors began with the hypothesis (see Chapter 17) that this conversion is similar to crystallization and results from an interaction between the normal prion and the ends of long fibers composed of pathogenic prion. In this process, conversion would involve a lengthening of these fibers at their extremities; its speed would thus be limited by the number of such fibers. On that basis, Soto's team believed that they could speed up the reaction by increasing the number of extremities. To do this, they subjected the mixture to ultrasound in order to break up the fibers. More precisely, their experiment alternated periods of incubation (enabling the fibers to grow longer with the addition of normal protein to their ends) with periods of ultrasound treatment. In one of the experiments they describe, in which a small quantity of protease-resistant pathogenic prion was added to a brain tissue extract containing a great deal of normal protease-sensitive prion, the quantity of protease-resistant prion was increased about thirtyfold.

In diagnostic terms alone, this is most promising because it could enable us to detect very small quantities of pathogenic prion, through what amounted to an amplification. And in the theoretical sphere it provides a powerful argument in favor of the theory that the conversion of the normal prion into a protease-resistant prion occurs through a process of quasi-crystallization. Every scientist in this field is now waiting to hear whether protein that has been converted in vitro turns out to be infectious. Experiments are under way, but the results were not known as of July 2002. If the outcome were positive and if it were reproducible with purified proteins, this would provide considerable support for the prion theory.

As shown in the foregoing summaries, 2001 held great promise for the diagnosis of TSEs. Treatment and prevention also saw advances, which we discuss next.

TREATMENT AND PREVENTION

For years, Stanley Prusiner's laboratory has been searching for molecules that could serve as drugs to treat prion diseases. In August 2001 it announced some apparently promising results.[4] Researchers there had focused on molecules that could reach the brain—those able to cross the blood-brain barrier. Among these, drugs long used to treat malaria, such as quinacrine, proved to cause a rapid decrease in the quantity of prion in cultures of infected cells. Nothing we now know can explain this; nevertheless, testing on CJD patients quickly ensued. Hopes were raised by press reports that the early stages of treatment saw a considerable improvement in the condition of one of the patients in this test, a British woman with new variant CJD. But these hopes were dashed in early December when it was learned that the woman had died. But while the hope proved to be premature, so, undoubtedly, was the disappointment. Perhaps study should focus in particular on the effect of these molecules on patients at a less advanced stage of the disease.

When you think of treatment and prevention, you immediately think of the body's immune reaction and of antibodies; we could draw on these either for a vaccine or for treatment. Prusiner's team published some encouraging results in 2001: Antibodies against the normal prion, when added to a cell culture simultaneously with a quantity of pathogenic prion, prevented the pathogenic prion from multiplying.[5] The most likely hypothesis would be that these antibodies prevent interaction between pathogenic and normal prions and thus stop the conformation of the normal prion from changing. Another result from Aguzzi's laboratory provided the first indication that antibodies against the normal prion in animals could prevent the spread of the pathogenic prion.[6] The approach here was rather complicated: It involved transgenetically introducing a gene encoding an anti-prion antibody into an inbred line of mice. The transgenic mice were resistant to intraperitoneal injections of pathogenic prions; they had been protected by the antibody. Moreover, they showed no sign of autoimmune disease—which, as we have noted, was a source of concern if an animal were to produce an antibody against a molecule found on the surface of many of its cells. These two outcomes are promising, although we are far from discovering a vaccine or an antibody-based treatment. Developing a vaccine remains especially problematic, because the prion—a protein belonging to the "self"—triggers no detectable immune reaction.

HAVE SHEEP BEEN CONTAMINATED?

There was much talk of sheep in 2001. At least two questions were being asked. The first and most urgent was whether they could have been contaminated by the BSE agent and whether sheep products could thus be capable of transmitting BSE to humans. The second, not entirely unrelated, question was whether the BSE epidemic could have started with the transmission of scrapie to cattle through the consumption of meat and bone meal (MBM).

In Chapter 23 we discussed the possibility that the BSE agent could be transmitted to sheep. We know that this can take place under experimental conditions, but not whether it has actually taken place in the British flock. The only existing way to know would be to gather a great number of brain samples from sheep with scrapie and use them to inoculate mice. Studying the incubation period along with the type and distribution of the lesions would make it possible to recognize the BSE agent. And such a study is indeed under way in the United Kingdom. In the summer of 2001, rumors began to circulate to the effect that the BSE agent had been found in sheep; the official outcome was to be announced at the end of the year. Europe's health authorities were in a state of red alert. If the results were positive, drastic steps would have to be taken in the sheep-farming sector. Then, just two days before the outcome was made public, there was a dramatic announcement: The researchers had made a mistake. They had mingled samples of sheep brains with samples of cattle brains—and thus there are still no data on the possible transmission of BSE to sheep in natural conditions.

If the BSE agent is indeed found in sheep, we will want to know if it had actually been transmitted from cattle though MBMs, or whether it had always existed in that species. Many strains of the scrapie agent are known to exist in sheep, and one of these—perhaps a rare strain—could have been transmitted to cattle, thus giving rise to BSE. This hypothesis will arise again later in this chapter.

Not knowing whether sheep have been contaminated, some scientists have tried to devise theoretical estimates. In a January 2002 article,

British epidemiologists estimated that at most a few dozen sheep alive at the time of their study had been contaminated by BSE.[7] Clearly that is a very small number in the context of the approximately forty million sheep in Britain. But the authors say that it could rise if effective horizontal transmission took root within flocks.

As these debates were under way, the British adopted a measure (and the French are following suit) that could eradicate scrapie among sheep flocks, irrespective of whether it was BSE in disguise. This was based on the observation (to which I referred in Chapter 18) that there is a natural polymorphism in the prion gene in sheep, and that certain forms of the gene entail resistance to scrapie. Just as the nature of the position-129 amino acid (methionine or valine) in humans plays a role in determining sensitivity to Creutzfeldt-Jakob disease, the nature of the amino acids at positions 136, 154, and 171 determines a sheep's sensitivity to scrapie.[8] The idea is to use appropriate selection methods to gradually strip the flock of the alleles of the gene that makes the animals sensitive to scrapie.

HOW *DID* THE COWS GO MAD?

Let us now return to the question that this book was supposed to answer: How *did* the cows go mad? More precisely, did this disease come about—as assumed at the beginning of the BSE epidemic—because cattle were fed meat and bone meal containing the causative agent of scrapie? The debate (of which I spoke in Chapter 21) is far from over. Why was an epizootic—an animal epidemic—declared at one particular time, the early 1980s, and only in the United Kingdom? Using animal-based meals to feed cattle is nothing new; it dates from the mid-nineteenth century and was first introduced in South America on the suggestion of Baron Justus von Liebig (who invented Liebig beef extract, among many other achievements). It then spread to other countries. The practice evolved during the 1970s throughout the industrialized world, in parallel with intensive animal raising. Changes in the methods by which MBMs were manufactured coincided with the emergence of BSE, but these were not a British monopoly; the changes also occurred in other countries, including France and the United States. And as for scrapie, it was already to be found throughout western Europe and the United States. It seemed as though, if BSE came from contamination of cattle by scrapie, such contamination should have taken place far earlier and in many countries.

That was the conclusion of a major British inquiry carried out under the direction of the judge Lord Phillips, whose report was made public in November 2000.[9] That extensive report maintained that the appearance of BSE was the result of a most unusual event. Such an event could have been the inclusion of the corpse of a cow affected by a particularly virulent form of spontaneous BSE in the manufacture of a batch of meat and bone meal. Another possible exceptional event that some researchers have mentioned concerns the death of six white tigers at the Bristol zoo between 1970 and 1977; they died of what was then diagnosed as a transmissible spongiform encephalopathy, but no one knows what became of the corpses. If they had been used to prepare meat and bone meal, tiger prion (possibly of exceptional virulence) could have been at the root of the BSE epidemic. Even though, looking back at it, the veterinarian who made the diagnosis is no longer certain that this was indeed a TSE, this rather exotic hypothesis illustrates how an exceptional event could have caused the epidemic. After all, it isn't often that a cow eats tiger in the way that we eat beef.

But the hypothesis that scrapie had been transmitted to cattle still has its supporters. Here we should mention the conclusions of a scientific review committee appointed by the British government and chaired by Professor Gabriel Horn of the University of Cambridge; its conclusions were issued in July 2001.[10] The committee found that this hypothesis could not be ruled out, and that a practice specific to the United Kingdom could have been at the root of BSE. This was the 1970s practice of feeding MBMs to calves, which could be more sensitive than adult animals to prion infection. In the committee's view, farmers in other European countries and in the United States hardly ever used MBMs to feed calves, but only to feed adult animals, especially lactating cows. If that is true, the hypothesis has a certain appeal. The belief that young animals are more prion-sensitive than adults is of very long standing (see Chapters 1 and 8), although it has never been demonstrated through rigorous experiments. In any case, because of the differences in properties between the BSE agent and all known strains of scrapie, this contamination had to have been caused either by a rare strain of the scrapie agent or by a strain that had been transformed by the processes used in manufacturing meat and bone meal.

So, the debate on the origins of BSE is not over, and it may never be. But although we do not know where BSE comes from, we can at least study the epidemic and try to predict its future course.

WHERE DOES THE BSE EPIDEMIC STAND?

In the United Kingdom, the BSE epidemic continues its decline. In 2001, the number of new cases was around fifty per month, whereas it had been double that figure in 2000. That decline may not be as speedy as one might have hoped, but that is surely due in part to ever more active monitoring. The thorniest point is that a small number of the cases declared in 2001 were in cattle born after 1996, the year that saw the adoption of the strictest prohibitions on feeding MBMs to farm animals, something we shall return to later. By the end of 2001, the total number of confirmed BSE cases in the United Kingdom since the beginning of the epidemic had risen to a little less than 180,000. Although cases had been reported in many other countries, they were far fewer in number.

In France, nearly five hundred confirmed cases had been reported by the end of 2001. Comparing the figures for 2000 and 2001 could be cause for concern, because they would seem to indicate that the epidemic is on the rise there: The number of confirmed cases rose from 162 in 2000 to 274 in 2001. But in fact this increase is entirely due to the gradual introduction, beginning in late 2000, of active screening, including systematic screening at slaughterhouses for animals aged thirty months or more. Taking account only of cases detected in animals with clinical symptoms of the disease—which was in effect the only kind of screening in place before mid-2000—the number fell from 102 in 2000 to 91 in 2001. Nothing here suggests that the epidemic is on the rise in France. Quite the contrary.

Another element that appears disturbing but that must be put in context is the number of countries that have been affected. This number was twenty by the end of 2001, up from eight in 1999, when the countries most heavily affected were the United Kingdom, Portugal, France, Ireland, and Switzerland. The detection in late 2000 of seven cases in Germany contributed to a Europe-wide panic; that country declared 125 cases in 2001. And in the same year several other European countries joined the club, some of them declaring a significant number of cases. These included Spain (with eighty-two cases), Italy (with forty-eight), and others with fewer than five, such as Austria, Finland, Greece, Slovakia, Slovenia, and the Czech Republic. The late-2001 announcement of three cases in Japan came as a particular surprise and had a significant effect on the Japanese cattle industry. What are we to think of this apparent spread to a growing number of countries? It is too early to say, but a decisive factor is probably intensified monitoring procedures. Otherwise it is hard to understand how a country like Spain or Italy could have been spared when contaminated British meat and bone meal must have been imported during the 1980s. In other words, there must have been cases of mad cow disease in these countries before 2001, but they had remained undetected.

Where the number of cases is very low, as in Japan, the origin of these cases raises a fundamental question—a question that in a few years could also arise in countries now affected by the epidemic but in which it is currently in decline: Do these occasional cases result from contamination, or are they sporadic, like cases of sporadic Creutzfeldt-Jakob disease in humans?

What is happening, finally, in countries thought to be free of BSE—the United States, for example? According to U.S. authorities, not one case has been detected, and very strict measures have been adopted to prevent the importation of animals or animal products from the United Kingdom or other countries affected by BSE. It is impossible, of course, to rule out occasional sporadic cases that might never be detected, but these should not give rise to an epidemic, because since 1997 MBMs have been banned in feed for ruminants. It seems likely, therefore, that the United States will not experience a mad cow crisis. Yet, there remains a nagging worry: the existence in the United States of chronic wasting disease, a transmissible spongiform encephalopathy affecting elk (see Chapter 22).[11] It was first identified in 1960 among animals in captivity, and it is now found among wild elk in quite extensive areas of northern Colorado, southeastern Wyoming, and Saskatchewan province. Nothing is known about its origin. It could have been initially transmitted to farmed elk through MBMs in their feed, but there is no proof of this. Transmission too is a mystery, but there is every indication that it could take place horizontally, as scrapie spreads within a flock. Indeed, how else can we explain its spread within wild populations? We do not know whether this disease can be transmitted to humans. As a precautionary measure, Colorado and Wyoming health authorities have enacted a number of measures intended to prevent hunters from eating meat from affected elk. The only really worrying scenario would be if horizontal transmission were to take place between elk and cattle. If the infectious agent then retained the capacity for horizontal transmission among cattle, the disease could become endemic, like scrapie among sheep.

THE HUMAN TOLL

What is the state of affairs with new variant Creutzfeldt-Jakob disease, which is almost certainly the result of the transmission of the BSE agent to humans? By the end of 2000 a total of eighty-six people had died of it: eighty-two in the United Kingdom, three in France, and one in Ireland. A year later the total was 101, meaning that there had been fifteen deaths from new variant CJD in 2001, about the same number as in 1998 (eighteen) and 1999 (fifteen), but far less than in 2000 (thirty). The epidemic was thus continuing, but to a limited extent. What else did we see in 2001? There was some indication of how cow-to-human transmission could have taken place, and new estimates—less pessimistic than those of the year before—were made concerning the future course of the epidemic and the probable number of new victims.

Two kinds of epidemiological data enabled scientists to offer hypotheses about the contamination of humans by BSE: data derived from the study of a group of cases centered in a central English village called Queniborough, and a broader study of dietary habits throughout the United Kingdom.[12]

The quiet little village of Queniborough, Leicestershire, was thrust onto the front pages when five young people from the vicinity, aged nineteen to twenty-four, died of new variant CJD between August 1998 and October 2000. These five accounted for about 5 percent of the total number of deaths from the disease in Britain. The probability that such a concentration of cases in so small an area could be coincidence was less than four in a thousand. This "cluster," as epidemiologists call it, had to be explained, and an inquiry was held. Its results were made public at the end of March 2001. It seemed that the five victims had all eaten meat from one of two butcher shops that used traditional methods of slaughter and butchering. Specifically, these butchers had themselves opened the animals' skulls to remove the brains, thus contaminating their tools. If an animal was affected by BSE, these tools could then transfer the infectious agent to many other cuts of meat. These methods could have led to the transmission of the infectious agent to a great number of humans. Although this might account for the Queniborough cases, it surely cannot extend to all cases in Britain, because the methods used by those two butchers had years earlier been abandoned by most other butchers in the United Kingdom and elsewhere in Europe. One interesting fact about the incubation period emerged from study of the Queniborough cases: Calculating on the basis of when the two butcher shops were closed, incubation would have been between ten and fifteen years, which more or less coincided with the average incubation period of kuru among the Fore people.

Other epidemiological data analyzed the distribution of new variant CJD cases in parallel with dietary habits. The incidence of the disease did not seem to be evenly distributed throughout Britain. The distribution of known cases at the end of 2000 indicated that incidence was twice as high in northern England and Scotland as in southeastern England and Wales, with intermediate figures for the central areas. Could these differences be related to differing dietary habits? A study of diet in various regions of Britain made it possible to correlate the incidence of CJD and the consumption of foods such as hamburgers, sausages, and meat pies, which most likely contain meat with a strong probability of having been contaminated by prion—so-called "mechanically recovered meat" removed from the spinal column and nervous system tissues. On the other hand, another study found no significant differences in the consumption of cattle products in the regions under consideration. Obviously, it is tempting to put our trust in the former study, which yielded the results we had expected. But the argument remains tenuous.

Thus, it is not yet possible to point the finger at any given dietary habit as the cause of infection by the BSE agent. Indeed, even though it remains the most likely hypothesis, there is no actual proof that contamination took place through food.

As we have seen, the number of deaths from new variant CJD rose only moderately in 2001. But clearly, that is not enough to enable us to conclude that the epidemic has peaked and will now begin a decline. This could be nothing more than a temporary respite to be followed by a sharp rise. Everyone wants to know what will actually happen—governments in particular, because the measures they adopt will depend on whether this is the precursor of a major epidemic affecting hundreds of thousands, or millions, of their citizens, or merely a flash epidemic that will claim only a few dozen or a few hundred victims. Therefore, since Roy Anderson's team offered its estimates in 2000, other epidemiologists have ventured to peer into the crystal ball.

In August 2000, Anderson's team estimated that the total number of victims in the United Kingdom could be anywhere between 100 and 136,000. As we saw in Chapter 22, the latter figure—although much trumpeted in the media—is based on the rather unlikely hypothesis that the average incubation period is sixty years. If that figure were reduced to twenty or thirty years, the maximal projected number of victims would not exceed a few thousand. Two newer analyses, published in November 2001, resulted in far less pessimistic estimates.[13] According to one of these, published by a Franco-British team, the maximal number of deaths should not exceed four hundred and could be closer to two hundred, or about double the present number. In order to arrive at that conclusion, the authors carried out an analysis taking account of the age of the victims. Since the very first cases, this has been notably low; of the ninety individuals who had died of nvCJD when the authors carried out their analysis, only six were older than fifty years of age. It appeared that, for unknown reasons, oral infection takes place far more easily in young individuals; this recalls what has often been noted in animals. The authors arrived at the figures cited above in light of that information, and assuming that most infection had taken place between 1980 and 1989. That is easy to understand. If the disease manifests itself most often before patients reach the age of thirty-five, the epidemic should end within about twenty years. The study suggests that 2000–2001 represented the height of the epidemic, and that the number of cases should decline beginning in 2002.

All of these estimates are based on theory and are thus uncorroborated. Still, they indicate that we should not fall prey to catastrophic pessimism, and that there are well-reasoned estimates by which the total number of deaths will not be in the millions or even thousands, but perhaps in the range of a few hundred. This does nothing to mitigate the tragedy of the disease for patients and their families, but it puts the public health dimensions of the problem in perspective.

Something that was not considered in developing these estimates, but which we ought to bear in mind, is the genetic characteristics of the affected individuals. We know that the nature of the prion gene's position-129 amino acid plays a major role in sensitivity to infection. The remarkable fact is that everyone affected by new variant CJD through the end of 2001 was M/M homozygous (see Chapter 20). If this means that only homozygotes are sensitive to the nvCJD agent, these estimates remain valid. But if that genetic configuration corresponds only to a lengthened incubation period, as seen in cases of contamination during human growth hormone therapy, the number of victims, which would then include both V/V homozygotes and M/V heterozygotes, could be about twice as high.

EPILOGUE

WE HAVE NOT YET BEATEN The Disease. By possessing many characteristics that run completely counter to orthodox scientific thinking, it has long managed to evade the pursuit mounted by veterinarians, physicians, and researchers.

The Disease is transmissible, but its incubation period is far longer than anything ever seen before. It is infectious, but it triggers no defensive reaction in the body—a strange characteristic that for years led researchers down the garden path. It is caused by an infectious agent that is resistant to all the usual decontamination processes; this is what inspired the heretical hypothesis of an infectious agent lacking nucleic acid: a prion composed exclusively of protein. Although it is infectious, The Disease may also be hereditary, and it can appear spontaneously—all in contravention of Pasteurian dogma.

Researchers have been forced to ascribe to the prion protein a number of properties that run counter to their normal thinking. It is a furtive protein that can cross the intestinal barrier, insinuate itself into the nervous system, crawl up to the brain, and enter the neurons and destroy them. Furthermore, this protein is thought to take on several different conformations—all of them stable and all affecting the disease's incubation period and the nature of the lesions it creates in the central nervous system. For a molecular biologist brought up to believe that for every protein there is one single three-dimensional structure, determined by its amino acid sequence, this is very difficult to accept.

Without a doubt, this strangeness of The Disease's characteristics and its ability to conceal itself through a variety of disguises delayed its identification. But there is another reason why the hunt has been a long one. Progress was intimately linked to developments in science and technology (see the Chronology at end of this book).

The achievements and delays of the hunt bring to the fore some aspects of scientific research that are unchanging although many people are unaware of them. One is the role of what we might call dogma, explanations of a natural phenomenon that are more or less universally accepted because they resulted from a series of consistent observations. An example is the role of nucleic acids as a foundation of heredity. Established on the basis of a sufficient quantity of experimental data, it has become dogma. It is not called into question every time a hereditary trait is studied; you must accept it as fact if you want to make any progress. That is why the notion that an infectious agent capable of reproduction could be without nucleic acid was inevitably met with the greatest skepticism for a long time. This kind of situation is by no means exceptional. Researchers often make observations that appear to contradict dominant theories; initially, it is the observations—or the researcher's interpretation of them—that are called into question. In order to persuade their colleagues that they were correct and that it is the dominant theories that ought to be questioned, researchers must be highly imaginative in presenting their results convincingly. In the course of the hunt for The Disease, it was necessary to upend—or *try* to upend—a whole body of dogma. This was certainly not conducive to rapid progress.

Another important aspect is the value of research on esoteric, apparently minor subjects, provided it is done properly. The trend today leans a little too much in the direction of carefully planned research focusing on subjects that, in the short or medium term, could have practical applications. If that approach had been strictly taken in the past, we would today be completely powerless in the face of the BSE epidemic and its public health consequences. We are fortunate that for years veterinarians had been studying scrapie, an obscure disease of sheep, and that physicians had been investigating Creutzfeldt-Jakob disease, an extremely rare condition. We are fortunate also that Gajdusek found the resources with which to study kuru, a strange disease affecting one tribe in Papua New Guinea, because if he had not, it would undoubtedly have taken a very long time to establish the link between animal and human TSEs.

In spite of all the media attention, the study of TSEs continues to be a relatively marginal activity. That is first of all because of its difficulty, but also because, quantitatively, these diseases are a minor public health problem. Sporadic Creutzfeldt-Jakob disease still affects only about one person in a million per year, and is thus the cause of just one death in ten thousand. And the total number of people affected by new variant CJD has been hardly more than a hundred. Of course, that number will inevitably rise. But even if we believe the most pessimistic forecasts, it should, in principle, remain very modest compared with the number of deaths from the other diseases that threaten us. Besides, this new disease ought to disappear in coming years in the wake of the measures taken to put an end to the BSE epidemic.

There could thus be a great temptation to avoid investing too lavishly in studying TSEs. That would be a serious mistake, because understanding their underlying mechanisms will bring fresh ideas to the entire field of biology. In addition, from the public health standpoint, a drug that can cure Creutzfeldt-Jakob disease, even if its profitability proved to be highly dubious, would be very likely to advance pharmaceutical research on other, far more common, degenerative diseases of the nervous system with obvious similarities to TSEs, such as Alzheimer's disease.

A final element that we can observe in the story of the hunt for The Disease is the compartmentalization that exists between fields of study. This is of enormous importance, but is rarely discussed. For evidence, we need only look at various recent episodes, including the mad cow crisis, in which lawyers, the press, and the public have cried, as with one voice, "They knew, and they did nothing!"

The ambiguity of that terse accusation lies in the word "they." Who knew what? It is a long time since any person could honestly claim to possess a substantial portion of the sum total of human knowledge. The days of Leonardo da Vinci are past. The extent of knowledge has become vast, and even the best-educated scientists cannot pretend to know more than an infinitesimal fragment of it. And however vast it is, it continues to grow. It is estimated today that an astonishing twenty-five million scientific papers are produced each year—about a hundred thousand a day.[1] Even limiting ourselves to the life sciences and to the most important publications as catalogued in major international databases, there are still some three hundred thousand articles a year, about a thousand a day. Even in the most circumscribed areas, it is extremely hard for researchers to stay up to date and still have time for their own work.

It is therefore not enough that work be published for it to be automatically known throughout the scientific community. A typical researcher has good knowledge about work in his own field, a little about related fields, and hardly any about more remote subjects. Take Gajdusek for example. When he began his work on kuru, the future Nobel laureate was completely unaware of the existence of Creutzfeldt-Jakob disease, even though it had been described forty years earlier, or of scrapie, which had been the subject of writings for two hundred years.

Here, let us return to the human growth hormone tragedy. Objectively speaking, as we have seen, there was every cause to be suspicious of such a therapy. But neither endocrinologists nor pediatricians had any reason to be aware of the veterinary literature on scrapie. Maybe "they" knew, but the particular people who were developing a treatment for pituitary dwarfism did *not* know.

Can similar situations be avoided in the future? Well, developments in electronic communications technology have facilitated access to an ever-growing body of information, and this ease of access will only increase with time. But in the final analysis, in order to obtain information you must look for it and have the time to digest it. Even if they had had access to the Internet, those endocrinologists and pediatricians were finding it hard enough to keep up with developments in their own fields, and would not have been seeking out articles about scrapie.

In the mid-eighteenth century, around the time English farmers were beginning to notice strange symptoms in their flocks, wordsmiths coined a term—"proteiform"—that today seems to be a doubly apt description of The Disease (although it was not used to describe it at the time). On the one hand, The Disease was able to elude its pursuers for so long by constantly changing its appearance, like the sea god Proteus. And on the other, it owed that amazing ability to the multiplicity of forms that a protein can take.[2]

In spite of its extraordinary properties, the proteiform Disease is only one among many examples of the new infectious diseases whose emergence was predicted by Louis Pasteur and Charles Nicolle, as quoted in the Prologue to this book. Mankind encountered quite a few of them toward the end of the millennium—AIDS, to name but one. All have been caused by infectious agents that were long held in a natural reservoir, but then were permitted to escape and spread among humankind because of changes in the way we live. According to Charles Nicolle, we are condemned never to be able to detect these new diseases from the outset. Yet the tale we have told proves that this is not precisely true. The appearance of iatrogenic Creutzfeldt-Jakob disease in children treated with human growth hormone was recognized remarkably soon after the appearance of the first cases in 1985. Similarly, new variant CJD, whose emergence had been feared in the wake of the BSE epidemic, was detected as soon as it appeared. Unfortunately, this turned out to be a disease with an extremely long incubation period, so that, both for iatrogenic disease and for BSE transmitted to humans, a considerable number of people had already been contaminated by the time the disease became apparent. Still, these diseases were indeed detected when they emerged, which bodes well for the future.

The growth of biological and medical knowledge, the refinement of diagnostic techniques, and the establishment of numerous epidemiological monitoring systems are enabling us with increasing frequency, and very promptly, to detect new infectious diseases. In the future, when we begin to take better account of interactions between humans and our environment, we may be able, perhaps, not only to detect new diseases, but to prevent them from coming into being.

NOTES

1: THE SHEEP ARE STRANGELY DIZZY

1. M. Mathieu, "Quelques mots sur la question ovine. Vente de béliers à Grignon," *Recueil de Médecine Vétérinaire* 53 (1876): 804–808.

2. T. Comber, *Real Improvements in Agriculture (on the Principles of A. Young, Esq.). Letters to Reade Peacock, Esq., and to Dr Hunter, Physician in York, Concerning the Rickets in Sheep* (London: Nicholl, 1772), 73–83.

3. *Journal of the House of Commons* 27 (1755): 87. See also T. Davis, *General View of the Agriculture of Wiltshire* (London: Phillips, 1811), 145–146.

4. I. Bertrand, H. Carré, and F. Lucam, "La tremblante du mouton," *Recueil de Médecine Vétérinaire* 113 (1937): 540–561.

5. Schmalz, "Observations relatives au rapport de M. Lezius, sur le vertige des moutons," *Bulletin des Sciences Agricoles et Économiques* 7 (1827): 217–219.

6. Roche-Lubin, "Mémoire pratique sur la maladie des bêtes à laine connue sous les noms de prurigo-lombaire, convulsive, trembleuse, tremblante, etc.," *Recueil de Médecine Vétérinaire* 25 (1848): 698–714.

7. L. Pasteur, "Sur la relation qui peut exister entre la forme cristalline et la composition chimique, et sur la cause de la polarisation rotatoire," *Comptes Rendus à l'Académie des Sciences* 26 (1848): 535–538.

2: MOLECULES AND MICROBES

1. L. Pasteur, *Examen critique d'un écrit posthume de Claude Bernard sur la fermentation* (Paris: Gauthier-Villars, 1879). Quoted in *Œuvres de Pasteur,* vol. 2 (Paris: Masson, 1922), 547. 9.

2. É. Duclaux, Pasteur: *Histoire d'un esprit* (Paris: Masson, 1896), 363.

3: MAD DOGS AND EARTHWORMS

1. É. Roux, "L'œuvre médicale de Pasteur," *Agenda du chimiste* (1896). Quoted by É. Duclaux in *Pasteur: Histoire d'un esprit* (Paris: Masson, 1896), 355–356.

2. E. Nocard, quoted by L. Nicol in *L'Épopée pastorienne et la médecine vétérinaire* (published by the author, 1974), 429.

3. H. Bouley, *Recueil de Médecine Vétérinaire* 51 (1874): 5.

4. F. Tabourin, "Sur la spontanéité et la contagion des maladies," *Recueil de Médecine Vétérinaire* 51 (1874): 263–291.

5. F. Tabourin, "Des générations dites *spontanées* et de leurs rapports avec les maladies parasitaires, infectieuses et virulentes," *Recueil de Médecine Vétérinaire* 55 (1878): 609–615.

4: SCRAPIE UNDER THE MICROSCOPE

1. J. Girard, "Notice sur quelques maladies peu connues des bêtes à laine," *Recueil de Médecine Vétérinaire* 7 (1830): 26–39.

2. C. Besnoit and C. Morel, "Note sur les lésions nerveuses de la tremblante de mouton," *Comptes Rendus de la Société de Biologie* 5 (1898): 536–538.

3. C. Besnoit, "La tremblante ou névrite périphérique enzootique du mouton," *Revue Vétérinaire* 24 (1899): 333–343.

4. S. Stockman, "Scrapie: An Obscure Disease of Sheep," *Journal of Comparative Pathology and Therapeutics* 26 (1913): 317–327.

5: CREUTZFELDT, JAKOB, AND OTHERS

1. H. G. Creutzfeldt, "Über eine eigenartige herdförmige Erkrankung des Zentralnervensystems," *Monographien aus dem Gesamtgebiete der Neurologie und Psychiatrie* 57 (1920): 1–18.

2. A. Jakob, "Über eine eigenartige Erkrankung des Zentralnervensystem mit bemerkenswertem anatomischen Befunde (Spastische Pseudosklarose-encephalomyelopathie mit disserminierten Degenerationsherden)," *Monographien aus dem Gesamtgebiete der Neurologie und Psychiatrie* 64 (1921): 147–228.

3. W. Spielmeyer, "Die histopathologische Forschung in der Psychiatrie," *Klinische Wochenschrift* 1 (1922): 1817–1819.

4. F. Katscher, "It's Jakob's Disease, not Creutzfeldt's," *Nature* 393 (1998): 11.

6: SCRAPIE IS INOCULABLE

1. J. Cuillé and P.-L. Chelle, "La maladie dite tremblante du mouton est-elle inoculable?" *Comptes Rendus à l'Académie des Sciences,* series D, 203 (1936): 1552–1554.

2. I. Bertrand, H. Carré, and F. Lucam, "La tremblante du mouton," *Recueil de Médecine Vétérinaire* 113 (1937): 586–603.

3. The cerebral lesions caused by the virus make sheep move oddly, with a bounding gait.

4. W.S. Gordon, "Advances in Veterinary Research. Louping Ill, Tick-Borne Fever and Scrapie," *Veterinary Records* 58 (1946): 516–520.

7: AND GOATS, *AND* MICE

1. R.L. Chandler, "Encephalopathy in Mice Produced by Inoculation with Scrapie Brain Material," *The Lancet* 1 (1961): 1378–1379.

8: SCRAPIE IS CONTAGIOUS

1. R.M. Ridley and H.F. Baker, *Fatal Protein: The Story of CJD, BSE, and Other Prion Diseases* (New York: Oxford University Press, 1998).

9: KURU AND THE FORE PEOPLE OF PAPUA NEW GUINEA

1. V. Zigas, *Laughing Death: The Untold Story of Kuru* (Clifton, N.J.: Humana Press, 1990), 1.

2. J.R. MacArthur, quoted in V. Zigas and D.C. Gajdusek, "Kuru: Clinical Study of a New Syndrome Resembling Paralysis Agitans in Natives of the Eastern Highlands of Australian New Guinea," *Medical Journal of Australia* 2 (1957): 745–754.

3. V. Zigas, *Laughing Death,* 133.

4. V. Zigas, *Laughing Death,* 142–144.

5. V. Zigas, *Laughing Death,* 226.

6. D.C. Gajdusek and V. Zigas, "Degenerative Disease of the Central Nervous System in New Guinea—The Endemic Occurrence of 'Kuru' in the Native Population," *New England Journal of Medicine* 257 (1957): 974–978.

10: THE WALL COMES DOWN

1. W.J. Hadlow, "Scrapie and Kuru," *The Lancet* 2 (1959): 289–290.

11: FROM PEARL NECKLACE TO DOUBLE HELIX

1. J.D. Watson, *The Double Helix,* ed. G.S. Stent (New York: W.W. Norton, 1981), among several editions.

2. J.D. Watson and F.H.C. Crick, "A Structure for Deoxyribose Nucleic Acid," *Nature* 171 (1953): 737–738.

12: THE PHANTOM VIRUS

1. T. Alper, D.A. Haig, and M.C. Clarke, "The Exceptionally Small Size of the Scrapie Agent," *Biochemical and Biophysical Research Communications* 22 (1966): 278–284.

2. T. Alper, W.A. Cramp, D.A. Haig, and M.C. Clarke, "Does the Agent of Scrapie Replicate without Nucleic Acid?" *Nature* 214 (1967): 764–766.

3. J.S. Griffith, "Self-Replication and Scrapie," *Nature* 215 (1967): 1043–1044.

13: A TRAGEDY IN THE MAKING

1. Endocrine glands are those that secrete within the body. They include the pituitary, the thyroid, and the pancreas, and the substances they secrete are dispersed in the blood or the lymph. Exocrine glands, such as the salivary and mammary glands, secrete their substances to the body's exterior.

2. M.S. Raben, "Preparation of Growth Hormone from the Pituitaries of Man and Monkey," *Science* 125 (1957): 883–884.

3. R.D.G. Milner, T. Russell-Fraser, C.D.G. Brook, et al., "The Experience with Human Growth Hormone in Great Britain: The Report of the MRC Working Party," *Clinical Endocrinology* 11 (1979): 15–38.

14: ONE CASE PER MILLION

1. C.L. Masters, J.O. Harris, D.C. Gajdusek, et al., "Creutzfeldt-Jakob Disease: Patterns of Worldwide Occurrence and the Significance of Familial and Sporadic Clustering," *Annals of Neurology* 5 (1979): 177–188.

15: PRIONS

1. S.B. Prusiner, "Novel Proteinaceous Infectious Particles Cause Scrapie," *Science* 216 (1982): 136–144.

2. Even under those conditions, the prion protein is not completely resistant to a given protease: A fragment of its chain, located at one of the two ends, is destroyed. But the rest, which remains intact, is where the protein's infectious properties lie.

16: APRIL 1985

1. P. Brown, "Human Growth Hormone Therapy and Creutzfeldt-Jakob Disease: A Drama in Three Acts," *Pediatrics* 81 (1988): 85–92.

2. Ibid.

3. G.A.H. Wells, A.C. Scott, C.T. Johnson, et al., "A Novel Progressive Spongiform Encephalopathy in Cattle," *Veterinary Records* 121 (1987): 419–420.

18: THE RETURN OF THE SPONTANEISTS

1. The dura mater is a relatively rigid membrane surrounding the brain. Fragments of this membrane are used in neurosurgery to protect small areas of the brain exposed during an operation. Like growth hormone treatments, such dura mater transplants resulted in the transmission of CJD to a comparatively large number of patients (140 as of July 2000), especially in Japan.

19: TO GROW—AND TO DIE

1. Montagnier's letter is annexed to the *Rapport sur l'hormone de croissance et la maladie de Creutzfeldt-Jakob,* published in December 1992 by the Inspection Générale des Affaires Sociales (IGAS, SA 19, No. 92145). Also annexed to the report is the transcript of the meeting of the administrative council of L'Association France Hypophyse cited in the chapter.

2. This spelling error indicates how unfamiliar CJD was to virologists at that time.

3. "Ban of Growth Hormone," *The Lancet* 1 (1985): 1172.

20: LESSONS LEARNED

1. P. Brown, C.D. Gajdusek, D.J. Gibbs, Jr., and D. Asher, "Potential Epidemic of Creutzfeldt-Jakob Disease from Human Growth Hormone Therapy," *New England Journal of Medicine* 313 (1985): 728–731.

21: HAVE THE COWS GONE MAD?

1. C. Siderius, *L'Alimentation des animaux domestiques, formulaires de rations* (Paris: Ballière, 1893), 30.

2. K.L. Morgan, K. Nicholas, M.J. Glover, and A.P. Hall, "A Questionnaire Survey of the Prevalence of Scrapie in Sheep in Britain," *Veterinary Records* 127 (1990): 373–376.

3. An article published in 1883 by a veterinarian from the Haute-Garonne named Sarradet ("Un cas de tremblante sur un boeuf," *Revue Vétérinaire* 31: 310–312) is sometimes advanced as proof that BSE already existed at that time. But, while they recall scrapie, the symptoms—notably pruritus, or itching, at the base of the tail—hardly coincide with those of BSE as we know it today. Moreover, the rapid development of the disease—just two weeks—is rather surprising. There is thus some question about what disease Sarradet really observed. It might have been a case of BSE with symptoms different from those of today—which would not be surprising, given the variability in the way transmissible spongiform encephalopathies are expressed in other species—but on the other hand, it might also have been a completely different disease.

22: FROM COWS TO HUMANS

1. T.A. Holt and J. Philips, "Bovine Spongiform Encephalopathy," *British Medical Journal* 296 (1988): 1581–1582.

2. D.M. Taylor, "Bovine Spongiform Encephalopathy and Human Health," *Veterinary Records* 125 (1989): 413–415.

3. Spongiform encephalopathy epidemics had been noted on several mink farms during the 1960s. The connection with scrapie was quickly made, and it was thought that the animals had been contaminated by an infectious agent in the meat they had been fed. At that time, scrapie was known only in sheep, and it initially seemed likely that they were the source of the agent. But later, epidemics were seen among mink that had never been fed sheep meat. These could have been contaminated by an agent with its origins in cattle, which would have suggested the existence of BSE in the United States, or by an agent from wild elk, which were also used to feed mink and which are subject to a disease similar to scrapie.

4. T.C. Britton, S. Al-Sarraj, C. Shaw, et al., "Sporadic Creutzfeldt-Jakob Disease in a 16-Year-Old in the UK," *The Lancet* 346 (1995): 1155.

5. R.G. Will, J.W. Ironside, M. Zeidler, et al., "A New Variant of Creutzfeldt-Jakob Disease in the UK," *The Lancet* 347 (1996): 921–925.

6. R.M. Anderson, C.A. Donally, N.M. Ferguson, et al., "Transmission Dynamics and Epidemiology of BSE in British Cattle," *Nature* 382 (1996): 779–788.

24: THE SECRET IN THE CLOSET

1. First of all, we have seen that the infectious protein was not completely resistant to proteinase K, which did indeed eliminate a small portion of the amino acid chain. The length of that portion is not exactly the same in all prion strains, which is reflected in differences in the size of the protein after protease treatment. Second, the prion protein is in fact a glycoprotein; sugars are linked to the amino acid chain at one or two points along that chain. The proportion of molecules that bear sugars at one or two points along the chain varies with the strain. Finally, differences have been observed in the interactions between prions of different strains and antibodies.

2. The normal protein used in these experiments was extracted from cells grown in the presence of an amino acid marked with a radioactive isotope and was thus itself radioactive. It was therefore possible to follow its progress after it had been mixed with the nonradioactive infectious protein, and to see that it had become resistant to proteinase K. Unfortunately, the change could be observed only in the presence of a large quantity of infectious protein, so that if the normal protein had become infectious the relative increase of the mixture's infectious strength would have been undetectably small.

3. An oligomer is a small polymer containing only a few monomers.

4. Results obtained subsequent to the publication of the French edition of this book and published in September 2001 support this second hypothesis. See K.J. Knaus, M. Morillas, W. Swietnicki, et al., "Crystal Structure of the Human Prion Protein Reveals a Mechanism for Oligomerisation," *Nature Structural Biology* 8 (2001): 770–774.

5. See note 2 above, and Chapter 27.

26: HAVE WE CONQUERED "THE DISEASE"?

1. In mid-December 2000, the Agence Française de Sécurité Sanitaire des Aliments (AFSSA) published the first results of a pilot BSE-screening program. Of the fifteen thousand cattle tested, 2.1 per thousand tested positive. But this was an at-risk sample, because it included only animals older than two years of age, which had been found dead or had been put down or slaughtered because of an accident, and which had come from the administrative districts most heavily affected by the epidemic. The percentage of contaminated animals among those that could actually enter the food chain is surely far lower. According to an article that appeared around the same time in the journal *Nature,* a maximum of about a hundred contaminated animals might have entered the food chain in France in the course of the year 2000. That might seem like a lot, but it means that only one animal in ten thousand killed for food would have carried the BSE agent.

27: 2001

1. M.B. Fischer, C. Roeckl, P. Parizek, et al., "Binding of Disease-Associated Prion Protein to Plasminogen," *Nature* 408 (2000): 479–483; and M. Mais-

sen, C. Roeckl, L. Glatze, et al., "Plasminogen Binds to Disease-Associated Prion Protein of Multiple Species," *The Lancet* 357 (2001), 2026–2028.

2. G.M. Shaked, Y. Shaked, Z. Kariv-Inbal, et al., "A Protease-Resistant Prion Protein Isoform Is Present in Urine of Animals and Humans Affected with Prion Diseases," *Journal of Biological Chemistry* 276 (2001): 31479–31482.

3. G.P. Saborio, B. Permanne, and C. Soto, "Sensitive Detection of Pathological Prion Protein by Cyclic Amplification of Protein Misfolding," *Nature* 411 (2001): 810–813.

4. C. Korth, B.C.H. May, F.E. Cohen, et al., "Acridine and Phenotiazine Derivatives as Pharmacotherapeutics for Prion Disease," *Proceedings of the National Academy of Sciences of the United States of America* 98 (2001): 9836–9841.

5. D. Peretz, R.A. Williamson, K. Kaneko, et al., "Antibodies Inhibit Prion Propagation and Clear Cultures of Prion Infectivity," *Nature* 412 (2001): 739–743.

6. F.L. Heppner, C. Musahl, I. Arrighi, et al., "Prevention of Scrapie Pathogenesis by Transgenic Expression of Anti-Prion Protein Antibodies," *Science* 294 (2001): 178–182; and F.L. Heppner, I. Arrighi, U. Kalinke, et al., "Immunity against Prions?" *Trends in Molecular Medicine* 7 (2001): 477–479.

7. R.R. Kao, M.B. Gravenor, M. Baylis, et al., "The Potential Size and Duration of an Epidemic of Bovine Spongiform Encephalopathy in British Sheep," *Science* 295 (2002): 332–335.

8. M.E.J. Woolhouse, P. Coen, L. Matthews, et al., "A Centuries-Long Epidemic of Scrapie in British Sheep?" *Trends in Microbiology* 9 (2001): 67–70.

9. "The BSE Inquiry: Report, evidence, and supporting papers of the inquiry into the emergence and identification of Bovine Spongiform Encephalopathy (BSE) and variant Creutzfeldt-Jakob Disease (vCJD) and the action taken in response to it up to 20 March 1996," available on the Internet at <http://www.bse.org.uk>.

10. "Review of the Origin of BSE," available on the Internet at <http://www.defra.gov.uk/animalh/bse/bseorigin.pdf>.

11. M. Enserink, "Is the U.S. Doing Enough to Prevent Mad Cow Disease?" *Science* 292 (2001): 1639–1641.

12. S. Cousens, P.G. Smith, H. Ward, et al., "Geographical Distribution of Variant Creutzfeldt-Jakob Disease in Great-Britain, 1994–2000," *The Lancet* 357 (2001): 1002–1005.

13. A.-J. Valleron, P.-Y. Boelle, R. Will, and J.-Y. Cesbron, "Estimation of Epidemic Size and Incubation Time Based on Age Characteristics of vCJD in the United Kingdom," *Science* 294 (2001): 1726–1728; and J.N. Huillard d'Aignaux, S.N. Cousens, and P.G. Smith, "Predictability of the UK Variant Creutzfeldt-Jakob Disease Epidemic," *Science* 294 (2001): 1729–1731.

EPILOGUE

1. The Science Citation Index (SCI) follows 4,500 journals, which publish a total of almost a million articles annually, or an average of about 200 to 225 articles per journal per year. The SCI indexes only the most important journals, however; the actual number of journals worldwide is estimated to be about 140,000. Applying the same formula to these 140,000 journals, we get a figure of twenty to thirty million. See M.-H. Magri and A. Solari, "L'évaluation au travers des revues scientifiques," in *Vie, Valeur et Valorisation de l'Information Scientifique* (Paris: Biotem Editions, 1997).

2. Curiously, the words "proteiform" and "protein" have quite different etymologies. The first, dating to 1761, comes from the name of the Greek god Proteus, while the second was coined in 1838 by the chemist Berzelius from the Greek word *prōteios,* meaning "primary," to designate the key component of living creatures.

BIBLIOGRAPHY

Baker, H. F., ed. *Molecular Pathology of the Prions*. Methods in Molecular Medicine. Totowa, N.J.: Humana Press, 2001.

Brown, P., and R. Bradley. "1755 and All That: A Historical Primer of Transmissible Spongiform Encephalopathy." *British Medical Journal* 317 (1998): 1688–1692.

Brown, P., M. Preece, J. P. Brandel, et al. "Iatrogenic Creutzfeldt-Jakob Disease at the Millennium." *Neurology* 55 (2000): 1075–1081.

Chastel, C. *Ces virus qui détruisent les hommes*. Paris: Éditions Ramsay, 1996, 252–284.

Collinge, J., and M. S. Palmer, eds. *Prion Diseases*. New York: Oxford University Press, 1997.

Court, L., and B. Dodet, eds. *Transmissible Subacute Spongiform Encephalopathies: Prion Diseases*. Paris: Elsevier, 1998.

Dealler, S. *Lethal Legacy: BSE—The Search for the Truth*. London: Bloomsbury, 1996.

Dubos, R. J. *Louis Pasteur: Free Lance of Science*. The Da Capo Series in Science. A Da Capo Paperback. New York: Da Capo Press, 1986.

Farquhar, J., and D. C. Gajdusek, eds. *Kuru. Early Letters and Field Notes from the Collection of D. Carleton Gajdusek*. New York: Raven Press, 1981.

Gajdusek, D. C. "Infectious Amyloids: Subacute Spongiform Encephalopathies as Transmissible Cerebral Amyloidoses." In *Fields Virology*, 3rd ed., edited by B.N. Fields, D.N. Knipe, P.M. Howley, et al. Philadelphia: Lippincott-Raven Publishers, 1996, 2851–2900.

Prusiner, S. B. "The Prion Diseases." *Scientific American* (January 1995), 30–37.

Prusiner, S.B. "Prions." In *Fields Virology*, 3rd ed., edited by B.N. Fields, D.N. Knipe, P.M. Howley, et al. Philadelphia: Lippincott-Raven Publishers, 1996, 2901–2950.

Prusiner, S.B., ed. *Prions Prions Prions*. Berlin and Heidelberg: Springer-Verlag, 1996.

Prusiner, S.B. "Prion Disease and the BSE Crisis." *Science* 278 (1997): 245–251.

Ratzan, S.C. *The Mad Cow Crisis, Health and the Public Good*. London: UCL Press, 1998.

Rhodes, R. *Deadly Feasts: Tracking the Secrets of a Terrifying New Plague*. New York: Simon and Schuster, 1997.

Ridley, R.M., and H.F. Baker. *Fatal Protein: The Story of CJD, BSE, and Other Prion Diseases*. New York: Oxford University Press, 1998.

Watson, J.D. *The Double Helix*. Edited by G.S. Stent. New York: W.W. Norton, 1981.

Zigas, V. *Laughing Death: The Untold Story of Kuru*. Clifton, N.J.: Humana Press, 1990.

CHRONOLOGY

DATE	THE HUNT	THE SCIENCE	THE DISEASE'S COUNTERATTACKS
MID-EIGHTEENTH CENTURY	Earliest known descriptions of scrapie are recorded.		
1848		Pasteur invents "three-dimensional chemistry."	
1860s		Mendel discovers what were later called genes.	
1870s		Pasteur and Koch demonstrate the role of microbes in contagious disease.	
1898	Besnoit discovers the presence of vacuoles in the neurons of sheep with scrapie.		
1900–1920		Science begins to address genetics and the chromosome theory of heredity.	
1918	Scrapie appears to be contagious in natural conditions.		
1920–1923	Creutzfeldt and Jakob describe the first cases of CJD.		
1931–1934			Gordon makes first attempts at a veterinary vaccine against louping ill.

DATE	THE HUNT	THE SCIENCE	THE DISEASE'S COUNTERATTACKS
1932		Fred Griffith identifies the transforming factor, which could change an inherited trait in a bacterium.	
1936–1938	Cuillé and Chelle demonstrate that scrapie is inoculable (transmissible).		
1937			Massive epidemic of scrapie occurs in Scotland among sheep vaccinated with one batch of louping ill vaccine.
1941		Beadle and Tatum develop the theory of one gene, one enzyme.	
1944		Avery and colleagues show that the trans-forming factor is made of DNA. Nucleic acids are the chemical basis of heredity.	
1953		Watson and Crick demonstrate the structure of DNA; beginnings of molecular biology.	
1955–1957	Zigas "discovers" kuru among the Fore people of Papua New Guinea.		
1957	Gajdusek joins Zigas. Klatzo notes the similarity between cerebral lesions in kuru and Creutzfeldt-Jakob disease.		Raben and others describe method of purifying human growth hormone from human pituitary glands.
1959	Hadlow notes the similarities between kuru and scrapie.		Doctors begin to treat pituitary dwarfism with growth hormone derived from human pituitaries.

DATE	THE HUNT	THE SCIENCE	THE DISEASE'S COUNTERATTACKS
1961	Pattison and Millson identify multiple strains of the scrapie agent. Chandler succeeds in transmitting scrapie to mice and performs the first quantification of the infectious agent.	Jacob and Monod study regulation of gene expression.	
1963	Gibbs and Gajdusek inoculate chimpanzees with brain extracts from Fore who had died of kuru.	Monod, Wyman, and Changeux study regulation of enzyme activity.	
1966	Gajdusek publishes findings on transmission of kuru to chimps. Alper and colleagues show that the scrapie agent is highly resistant to ionizing radiation and ultraviolet light.		
1967	How can a protein be infectious? J. Griffith's hypotheses describe two mechanisms.		
1968	Hypothesis that scrapie is contagious in sheep is confirmed. Like kuru, CJD is transmitted to chimpanzees. CJD and scrapie are grouped together as subacute spongiform encephalopathies (later called transmissible subacute spongiform encephalopathies, or TSSEs). British researchers identify a gene for sensitivity to scrapie infection in mice.		

DATE	THE HUNT	THE SCIENCE	THE DISEASE'S COUNTERATTACKS
1973			France begins production and distribution of human growth hormone.
1974			American woman contracts CJD following cornea transplant: first recognized iatrogenic transmission of CJD.
1975–1985		Researchers develop genetic engineering techniques (cloning and gene sequencing).	
1976	Nobel Prize is awarded to Gajdusek.		
1979	Annual incidence of CJD is about one case per million inhabitants in all countries. No geographic correlation between this and the incidence of scrapie in sheep.		
1980	Gajdusek reports transmission of kuru and CJD to squirrel monkeys through their feed.		Montagnier's letter to L'Association France Hypophyse warns of possible infection associated with human growth hormone treatment.
1980–1990		Development of nuclear magnetic resonance techniques and construction of high-resolution spectrometers make it possible to clarify the structure of proteins in solution.	
1982	Hadlow and colleagues find further evidence of natural contagion of scrapie, probably through oral route. The scrapie agent is believed to be completely or principally proteinaceous. Prusiner names it the prion.		

DATE	THE HUNT	THE SCIENCE	THE DISEASE'S COUNTERATTACKS
1985	Weissmann, Prusiner, and colleagues show that the prion gene is present in all mammals.		First cases of CJD are reported in young people treated with human growth hormone. First (retrospective) cases of BSE are identified in British cows.
1986–1988			Most countries begin gradual replacement of cadaveric human growth hormone with hormone obtained by genetic engineering.
1987			Wells and colleagues publish first scientific article on BSE.
1988			Wilesmith and colleagues demonstrate the role of meat and bone meal in the onset of the BSE epidemic; British government takes first measures to put an end to the epidemic.
1989	Scientists find first indication that inherited CJD results from prion gene mutations.		
1989–1990	Weissmann and Prusiner study the "kiss of death": The infectious prion can shape the normal prion in its own image.		
1990–1992		Molecular biologists develop techniques to selectively inactivate genes in mouse embryos.	
1991			Wilesmith and colleagues' epidemiological study suggests that BSE-agent transmission through meat and bone meal could result from halting the use of organic solvents in the manufacturing process.
1992–1993	Weissmann, Prusiner, and colleagues show that inactivation of the prion gene in mice results in resistance to the scrapie agent. The prion protein is thus a key component of the infectious agent.		

DATE	THE HUNT	THE SCIENCE	THE DISEASE'S COUNTERATTACKS
1993	Cohen, Prusiner, and colleagues note structural differences between the normal prion (rich in alpha-helices) and the infectious prion (high content of beta-sheets).		
1994	First arguments on the existence of structural differences among the agents of various scrapie strains appear in the literature. Researchers perform in vitro conversion of normal proteinase-K-sensitive protein into proteinase-K-resistant protein, like the infectious protein, by mixing it with the latter; but technically impossible to demonstrate whether this change is accompanied by the acquisition of infectious characteristics.		
1996			Will and colleagues identify new variant of CJD in ten young Britons, probably the result of eating cattle products containing the BSE agent.
1996–1997	Wüthrich, Glockshuber, and colleagues describe the three-dimensional structure of the prion protein in its normal form.		
1997	Nobel Prize is awarded to Prusiner.		Researchers demonstrate that the BSE and nvCJD agents have very similar properties, confirming the hypothesis that humans were contaminated by the BSE agent.

DATE	THE HUNT	THE SCIENCE	THE DISEASE'S COUNTERATTACKS
1997–2000	Researchers describe role of the lymphatic system (especially the spleen) in the transmission of the prion from the intestine to the brain.		
2000			Fear spreads regarding possible transmission of the nvCJD agent via blood transfusion.
			Anderson predicts a possible 130,000 UK deaths from nvCJD.
2001	Soto suggests that amplification of the proteinase-K-resistant prion in vitro could be the basis for a very sensitive diagnostic test.		New estimates (Valleron et al., Huillard d'Aignaux et al.) predict a maximum of a few hundred deaths from nvCJD.

ACKNOWLEDGMENTS

First of all, I apologize to all the many colleagues to whose work I have referred without attribution. By citing relatively few names, I sought to avoid weighing down the text with an excessively scholarly appearance.

My wife Brigitte strongly encouraged me to write this book, and was a constant source of support throughout the project. My many conversations with her, and with our daughter Barbara, were invaluable.

One colleague and very good friend, who played a major role in my decision to undertake this project, sadly died just as I was about to begin writing: Florian Horaud, who had written a study of this subject, which he kindly placed at my disposal. His premature death affected me deeply, and I wish to pay tribute to his memory.

Many colleagues lavished encouragement upon me, provided me with documents, read all or part of the manuscript, and suggested improvements. I would mention in particular Michel Brahic, Jean Castex, Michel Goldberg, Jean-Louis Guénet, Benno Muller-Hill, Michèle Mock and Agnès Ullmann. My old friend Franck Laloë twice read the text with a layman's eye and identified many passages that needed clarification. His interest in the project gave me reason to hope that I was not on the wrong track. Claude Chastel gave me access to the archives he had assembled on this subject for his book *Ces virus qui détruisent les hommes*. My thanks go to all these men and women.

I am grateful also to the staff of the library of the Institut Pasteur, and in particular to its director, Corinne Verry, as well as to Elisabeth Grison and Agnès Grébot, who, respectively, headed the libraries of the Ecole Nationale Vétérinaire d'Alfort and of the Agence Française de Sécurité Sanitaire des Aliments. They made it possible for me, with the help of my secretary, Yolande Meunier, to gather the bulk of the scientific articles and other works on which this book was based. Annick Perrot, curator of the Musée Pasteur, was an inexhaustible source of information on Pasteur and his followers. I experienced a warm welcome at the Rural History Centre of the University of Reading, United Kingdom, where I found a number of old documents on scrapie. Finally, I am very grateful to the late Dr. Charles Mérieux and to Betty Dodet, both of whom made it possible for me to participate in a November 2000 symposium on prion diseases. There I was able to meet the leaders in the field and ask them the questions that were on my mind.

Finally, in preparing Chapter 27, which outlines the events of 2001, I profited from information and advice from Annick Alpérovitch, Adriano Aguzzi, Bruce Chesebro, Dominique Dormont, Claudio Soto, Charlie Weissmann, Robert Will, and a number of individuals from the Agence Française de Sécurité Sanitaire des Aliments (AFSSA). My warm thanks go to them all.

INDEX

Adapters, and reading DNA, 79
Aggregates, 101, 138
Aguzzi, Adriano, 185
Alpers, Michael: epidemiological study of
    kuru by, 68–69; kuru transmitted to
    chimpanzees by, 69
Alper, Tikvah, 83–84
Alzheimer's disease: CJD brain lesions
    similar to those in, 37; kuru plaques
    similar to those in, 64
America. *See* Canada; United States
Amyloid plaques: among CJD victims,
    140; among nvCJD victims, 158–159;
    among GSS patients, 120; prions and,
    102; similarities of in kuru, CJD and
    scrapie, 102
Anderson, Gray, 62
Anderson, Roy, 160–161, 181, 196
Animals: failure to transmit BSE, 165.
    See also specific animals
Anthrax: artificial fertilizer and trans-
    mission of, 19–20; Pasteur's work
    with, 15–16; theories on transmission
    of, 18–20
Antibodies: against prions, 102; lack of in
    scrapie, kuru or CJD, 82; treatments
    and, 187
Anxiety, as a first symptom of CJD, 93
Artificial fertilizer, and anthrax
    transmission, 19–20

L'Association France Hypophyse:
    concerns about HGH treatment, 129;
    treating dwarfism and, 91
Asymmetry of molecules, 13–14
Australia: arrival of scrapie in, 45;
    incidence of CJD in, 97
Autosuggestion, and kuru, 62
Avery, Oswald, 77

Bacillus anthracis, 15. *See also* Anthrax
Backer family cases of CJD, 35, 36
Baker, Harry, 52–53
Base pairs, in genetics, 79
Beadle, George, 76
Bertrand, Ivan: attempt to inoculate
    sheep with scrapie, 39–40; scrapie
    symptoms described by, 7
Besnoit, Charles: discovery of scrapie
    lesions, 24, 25; scrapie as infectious
    and, 25–26
Blood: contagion of BSE and, 166–167;
    French restrictions on blood donations,
    167; as poor medium for inoculation
    of scrapie, 40; scrapie "virus" not
    detected in, 46; transfusions and
    possible transmission of nvCJD,
    166
Blumberg, Baruch S. (Nobel laureate),
    71
Bouley, Henri, 20–21

Designer: Victoria Kuskowski
Compositor: Michael Bass & Associates
Text: 11/15 Granjon
Display: Frutiger
Printer and Binder: Maple-Vail Book Manufacturing Group